Environmental Aspects, Applications, and Implications of Nanomaterials and Nanotechnology

Topical Conference at the 2011 AIChE Annual Meeting

Minneapolis, Minnesota, USA
16-21 October 2011

ISBN: 978-1-61839-583-2

Printed from e-media with permission by:

Curran Associates, Inc.
57 Morehouse Lane
Red Hook, NY 12571

Some format issues inherent in the e-media version may also appear in this print version.

Printed by Curran Associates, Inc. (2012)

For permission requests, please contact AIChE
at the address below.

AIChE
3 Park Avenue
New York, NY 10016-5991

Phone: (203) 702-7660
Fax: (203) 775-5177

www.aiche.org

Additional copies of this publication are available from:

Curran Associates, Inc.
57 Morehouse Lane
Red Hook, NY 12571 USA
Phone: 845-758-0400
Fax: 845-758-2634
Email: curran@proceedings.com
Web: www.proceedings.com

TABLE OF CONTENTS

(217a) Assessment of Potential Exposure to Carbon Nanotubes In the Production of Polymer Nanocomposites

Drew J. Thompson[1], Lin Li[1], Jing Wang[2] and David Y.H. Pui[1], (1)Mechanical Engineering, University of Minnesota, Minneapolis, MN, (2)Institute of Environmental Engineering, ETH Zurich, Switzerland

Recent animal toxicology studies have shown that carbon nanotubes (CNTs) may pose a significant health risk to those exposed to this nanomaterial in the workplace. However, many questions regarding the toxicology of CNTs still remain unanswered. There has yet to be agreement on which dose metric would not only correlate with the negative health effects but also be able to be successfully implemented in field measurements. The roles of the inclusion of metal catalyst particles in and the morphology of CNTs and/or CNT structures in the toxicology of CNTs are still being investigated. While these matters are being researched further by toxicologists there remains a dearth of information on the actual exposure to CNTs. Even if CNTs are found to be hazardous, there will be no risk without exposure. To fully understand this potential risk effort must be taken to measure the occupational exposure to CNTs.

In this study the potential occupational exposure to CNTs in the production of polymer nanocomposites, where CNTs were used as a filler material, was assessed. The potential exposure to CNTs was quantified by analyzing respirable dust samples collected by personal sampling for elemental carbon. The mass concentration of elemental carbon, of which CNTs mainly consist, was determined by thermal-optical analysis of the sample. Potential exposure sources were identified and characterized by both real-time measurements and area sampling. The potential sources investigated were comprised of both production equipment and production related tasks, which included the handling of unbound CNTs. The real-time measurements were performed by instrumentation which provided number concentration, surface area concentration, respirable mass concentration, and size distributions for micron-sized particles. Area samples were analyzed by transmission electron microscopy to characterize the morphology and to calculate the number concentration of CNT structures. Other samples were taken for analysis by inductively coupled plasma mass spectrometry to measure the concentration of trace metals. The concentration of these metals was then used to estimate the CNT mass concentration based on known impurities. Thermal-optical analysis was performed on additional area samples to measure the elemental carbon mass concentration.

See more of this Session: Environmental Health and Safety (EHS) Concerns of Nanomaterials
See more of this Group/Topical: Topical C: Environmental Aspects, Applications, and Implications of Nanomaterials and Nanotechnology

(217b) A Novel Technique for In-Vivo Toxicological Characterization of Engineered Nanomaterials

Georgios A. Sotiriou[1], Edgar Diaz[2], Mark Long[2], John Godleski[2], Joseph Brain[2], Sotiris E. Pratsinis[1] and Philip Demokritou[2], (1)Department of Mechanical and Process Engineering, Particle Technology Laboratory, ETH Zurich, Zurich, Switzerland, (2)Department of Environmental Health, Harvard School of Public Health, Boston, MA

How do we balance the potential hazards from new and inadequately characterized engineered nanomaterials with the potential and power of nanotechnology? It has become apparent, that potential environmental health effects are critical factors in the likely success or failure of new nanotechnologies. The current nanotoxicology approach of evaluating the toxic properties of the large universe of existing and emerging ENMs at one material at the time, places a significant financial burden to an emerging industry and may slow down significantly the development of this emerging industry. Alternatively, a promising approach will be to examine families of engineered and rigorously characterized particles, and to study the role of such factors as particle size, composition, shape, charge and surface chemistry and their link to specific biological outcomes.

Here, a novel method is explored which is suitable for ENM *in-vivo* inhalation and *in-vitro* toxicological characterization studies. ENMs are produced continuously in the gas phase using industry relevant flame spray pyrolysis reactors, allowing their continuous transfer to inhalation chambers, without altering their state of agglomeration. Important properties of the generated aerosols (i.e. particle size, concentration, shape, state of agglomeration, surface chemistry) can be modified allowing for both *in-vitro* and *in-vivo* investigations of toxicity. The ability of the developed technique to generate a variety of industry relevant, property controlled exposure atmospheres for inhalation studies was systematically investigated. The suitability of the technique to characterize the toxicity of inhaled ENMs in intact animal models was also shown in an *in-vivo* study involving Sprague-Dawley rats, using freshly generated nano-iron oxide (Fe_2O_3) as a test aerosol. We demonstrated both pulmonary and systemic toxicity using a variety of toxicological assays under acute exposure conditions. The future use of this novel method will improve our understanding in terms of the health and safety of ENMs and help us to investigate further our central hypothesis that physical and chemical characteristics of ENMs determine their bioavailability, redistribution, and toxicity in the lungs.

See more of this Session: Environmental Health and Safety (EHS) Concerns of Nanomaterials
See more of this Group/Topical: Topical C: Environmental Aspects, Applications, and Implications of Nanomaterials and Nanotechnology

(217c) Evaluation of the Toxicity of Nanomaterials Based On Knowledge Extraction From High Throughput Screening of Biological Toxicity Data

Rong Liu[1], Saji George[1], Bryan France[1], Robert Damoiseaux[1], Robert Rallo[2], Kenneth Bradley[1], Andre E. Nel[3] and Yoram Cohen[4], (1)California NanoSystems Institute, Center for the Environmental Implications of Nanotechnology (CEIN), Los Angeles, CA, (2)Enginyeria Informatica i Matematiques, Universitat Rovira Virgili, Tarragona, Spain, (3)Division of NanoMedicine, UCLA, Los Angeles, (4) Chemical & Biomolecular Engineering, University of California, Los Angeles, Los Angeles, CA

Nano-sized materials are increasingly utilized in many industrial products and processes, with over 1000 commercial products utilizing engineered nanomaterials (eNMs), primarily due to their unique nano-scale properties. Despite the many beneficial uses of nanomaterials, there is a growing public concern regarding the potential release of eNMs to the environment and the potential adverse impacts of exposures to eNMs. Although there are mounting studies on the toxicity of eNMs, understanding of the general principles governing eNMs toxicity potential and the long-term environmental health and safety associated with eNM containing products is in its infancy. In this regard, high throughput toxicity screening is critical for characterization of the potential hazard of eNMs, which is in turn indispensable information for subsequent risk assessment and development of environmental and health regulatory policies. In the present work an approach to knowledge extraction from high throughput screening (HTS) of nanoparticle toxicity is presented, based on RAW and BEAS-2B mammalian cell lines, focusing on identification of toxicity outcomes (i.e., "hit identification"), development of predictive nano-quantitative-structure-activity relations (nano-SARs), identification of pathway linkages and correlation of cell signaling pathways and cytotoxicity. Based on newly developed feature selection approach, a classification based nano-SAR was developed for cytotoxicity of metal and metal oxide nanoparticles. The approach was demonstrated for a library of metal and metal oxide nanoparticles to develop a classification based nano-SAR that can be used to identify decision boundaries with respect to hazard ranking of nanoparticles. In addition, self-organizing maps (SOM) analysis was performed to identify clusters corresponding to sub-lethal pro-inflammatory responses associated with ROS generation, lethal genotoxic responses, and cytotoxicity. Finally, complex network theory methods were applied to identify relationships between cell responses (pathway activation and cytotoxicity measures) as well as physicochemical properties of nanoparticles. For example, the analysis based on the above cell lines revealed a hierarchical activity-activity pattern where sub-lethal effects such as ROS generation and the intracellular Ca^{2+} flux are strongly related to lethal effects (e.g. cell membrane damage and cell death) and that eNM primary size, aggregation and its dissolution tendency were critical parameters affecting the observed toxicity behavior.

(217d) Human Blood Biocompatibility of Silver Nanoparticles

Jonghoon Choi[1], Vytas Reipa[2], Victoria M. Hitchins[3], Peter L. Goering[3] and Richard A. Malinauskas[3], (1)Department of Chemical Engineering, The David H. Koch Institute for Integrative Cancer Research, Massachusetts Institute of Technology, Cambridge, MA, (2)Biochemical Science, National Institute of Standards and Technology, Gaithersburg, MD, (3)Center for Devices and Radiological Health, Food and Drug Administration, Silver Spring, MD

Silver nanoparticles are increasingly being used in medical devices for their antimicrobial properties. Present toxicological techniques may not be sufficient to fully determine their safety for use in medical devices. In this talk, *in vitro* assessment of the hemolytic properties of silver particles in human blood will be discussed. In addition to the observed hemolysis, the possible mechanisms of red blood cell damage were explored in regard to the physical and chemical properties of silver particles. Four silver particle (two nano-sized, two micro-sized) preparations were dispersed in water and characterized using multiple analytical methods. Particle size and agglomeration were dependent on the suspension media. Under similar conditions to the hemolysis assay, i.e., with particles added to phosphate buffered saline and plasma, the size of the nanoparticles increased compared to particles suspended in water. Nano-sized particles exhibited higher hemolysis than micro-sized particles at mass concentrations greater than 220 ug/ml. While significant particle size changes occurred due to interaction with media components, the higher level of hemolysis observed with nanoparticles compared to micro-sized particles may be related to greater surface area, increased silver ion release, and direct nanoparticle interactions with red blood cells. These *in vitro* hemolysis results suggest that blood-contacting medical devices containing silver nanoparticles should be carefully evaluated for their hemolytic potential.

See more of this Session: Environmental Health and Safety (EHS) Concerns of Nanomaterials
See more of this Group/Topical: Topical C: Environmental Aspects, Applications, and Implications of Nanomaterials and Nanotechnology

(217e) Nanotechnology - What's So Big about the Small Stuff?

William C. Looney, Environment, AECOM, Milwaukee, WI

Nanotechnology - What's So Big about the Small Stuff?

Nanotechnology is "the Next Big Thing." It is a truly internationalphenomenon that "will have a major impact on the health, wealth and security ofthe world's people that will be at least as significant in this century asantibiotics, the integrated circuit, and manmade polymers." [1]

The market for nanotechnology-based products is expected to reach $2.5trillion by 2015, up from $147 billion in 2007.[2]

Nanotechnologypresents the promise of a diverse array of manufactured goods and products thatincorporate improved and innovative properties. Studies have determined that, due to their unique properties, nanomaterials (NM) have unpredictableproperties in biological systems and that the health implications from NMexposures are not yet known. Although numerous productsincorporating nanotechnology are here today, specific regulatory drivers for NM are still emerging. Companies producing or using NM must addressthe potential hazards of NM in the workplace to comply with existing workplace standards under the Occupational Healthand Safety Act. For example, the GeneralDuty Clause requires that employers provide each employee a place of employmentthat is free from recognized hazards that are causing or are likely to causedeath or serious physical harm and the Hazard Communication Standard (29 CFR1900.1200) provides workers with the right to know the hazards that they areexposed to in the workplace. Uncertaintyabout adverse health effects from exposures to NM and which work practices toadopt today should not impede efforts by companies to develop worker safetyprograms that minimize potential health effects. Such an approach can reduce the likelihood ofworker exposures to NM and the financial costs of subsequent liability claimsand litigation.

This presentation will provide an overview ofnanotechnology basics, an inventory of potentially applicable regulations andguidelines and the benefits of a proactive approach.

[1]National Science and Technology CouncilCommittee on Technology

[2]*Lux Populi Newsletter*, February 22, 2010, LuxResearch

Contact Information:

William C. Looney

Senior Program Manager

Director AECOM Nanotechnology Initiative

AECOM Environment

D 414.577.1362   C 262.893.0658

bill.looney@aecom.com

AECOM

11425 W. Lake Park Drive, Suite 100

Milwaukee, WI 53224

T 414.359.3030 F 414.359.0822

www.aecom.com

See more of this Session: Environmental Health and Safety (EHS) Concerns of Nanomaterials
See more of this Group/Topical: Topical C: Environmental Aspects, Applications, and Implications of Nanomaterials and Nanotechnology

(217f) Forensic Analysis of Nano-Material Environmental Regulation

Coby A. Scher and Dennis L. Caputo, Quest Consulting, Inc., Bellaire, TX

Forensic Analysis of Nanomaterial Environmental Regulation

Coby A. Scher – Quest Consulting, Inc. (Presenter)

& Dennis L. Caputo – Quest Consulting, Inc.

Some considerable effort has recently been displayed by manufacturers, researchers, regulatory bodies, professional organizations, NGOs, and other interested parties addressing how nanomaterials and their manufacture should be environmentally regulated. In the USA, everyone including EPA, FDA, ABA, universities and a host of other entities (including companies) are trying to voice an opinion. The executive branch of the federal government with congress's approval and funding has created an investigatory group examining the regulatory potential. Internationally, the EU and ISO are investigating the subject of "nano-regulation". There is a body of thought within science and engineering termed forensics. In fact, the pure definition of forensic according to Webster's Collegiate Dictionary, 10th edition, is "1; belonging to, used in, or suitable to courts of judicature or to public discussion and debate 2: Argumentative, rhetorical 3: Relating to or dealing with the application of scientific knowledge to legal problems." All three parts of the definition are applicable to the issue at hand – the application of forensics to the environmental regulation of nanomaterials. It is the authors' contention that use of environmental forensics to encompass scientific investigative methods will provide the answer to what level of regulation is required. Forensic environmental science and engineering is a state of mind; an analytical approach to solving this problem. And it need not be complicated, just complete and scientifically defensible. This paper will update the current governmental movement toward environmental regulation of nanomaterials and apply a forensic approach to solving the problem of nanomaterial environmental regulation.

See more of this Session: Environmental Health and Safety (EHS) Concerns of Nanomaterials
See more of this Group/Topical: Topical C: Environmental Aspects, Applications, and Implications of Nanomaterials and Nanotechnology

Effect of Shear Stress on Cytotoxicity of Silica Nanoparticles

Donghyuk Kim, Yu-Shen Lin, and Christy L. Haynes

In this work, nanotoxicity in the bloodstream was modeled as the cytotoxicity of sub-50 nm mesoporous silica nanoparticles to human endothelial cells was investigated under microfluidic flow conditions. Compared to traditional *in vitro* cytotoxicity assays performed under static conditions, unmodified mesoporous silica nanoparticles show higher and shear stress-dependent toxicity to endothelial cells under flow conditions. Interestingly, even under flow conditions, highly organo-modified mesoporous silica nanoparticles show no significant toxicity to endothelial cells. This paper clearly demonstrates that shear stress is an important factor to be considered in *in vitro* nanotoxicology assessments and provides a simple device for pursuing this consideration.

(292b) Quantifying the Origin of Nanosilver Ions and Their Antibacterial Activity

Georgios A. Sotiriou[1], Andreas Meyer[2], Jesper Knijnenburg[1], Sven Panke[2] and Sotiris E. Pratsinis[1], (1) Department of Mechanical and Process Engineering, Particle Technology Laboratory, ETH Zurich, Zurich, Switzerland, (2)Department of Biosystems Science and Engineering, ETH Zurich

Nanosilver is one of the most prominent components in several nanotechnology products, primarily for its bactericidal and plasmonic properties [1,2]. Its use, however, is questioned as U.S. EPA has been petitioned to label it as pesticide. To better understand its antibacterial activity, nanosilver with closely controlled size immobilized on nanostructured silica is generated by flame aerosol technology [3]. The antibacterial activity of small (<10 nm) nanosilver particles is attributed mostly to released Ag^+ ions from their surface upon contact with water [1]. The origin of these ions is explored systematically now by conditioning the surface of various nanosilver particles (4-9 nm) immobilized on nanostructured silica by wet-impregnation and dry flame-spray pyrolysis. Silica prevents nanosilver flocculation without the need of any surface functionalization (e.g. citrate) that would compromise the Ag^+ ion release kinetics. Washing or reducing the Ag/SiO_2 nanoparticles removes the oxide layer from the nanosilver surface minimizing their ion leaching. The amount of leached Ag^+ ions corresponds to the dissolution of one or two silver oxide monolayers. Finally, the antibacterial activity of small (< 10 nm) nanosilver particles that have their surface oxide layer removed or reduced is significantly lower than the freshly prepared ones by either wet or dry methods.

1. Sotiriou, G. A.; Pratsinis, S. E., Antibacterial activity of nanosilver ions and particles. *Environ. Sci. Technol.* **2010,** *44*, 5649.

2. Sotiriou, G. A.; Sannomiya, T.; Teleki, A.; Krumeich, F.; Vörös, J.; Pratsinis, S. E., Non-Toxic Dry-Coated Nanosilver for Plasmonic Biosensors. *Adv. Funct. Mater.* **2010,** *20*, 4250.

3. Sotiriou, G. A.; Teleki, A.; Camenzind, A.; Krumeich, F.; Meyer, A.; Panke, S.; Pratsinis, S. E. Nanosilver on nanostructured silica: Antibacterial activity and Ag surface area. *Chem. Eng. J.* **2011,** *170*, 547.

See more of this Session: Environmental Implications of Nanomaterials: Biological Interactions
See more of this Group/Topical: Topical C: Environmental Aspects, Applications, and Implications of Nanomaterials and Nanotechnology

(292c) Oxidation of Biomolecules by Emerging Inorganic Nanoparticles

Antonia Luna-Velasco[1], Reyes Sierra-Alvarez[2] and Jim A. Field[2], (1)Chemical and Environmental Engineering, University of Arizona, Tucson, AZ, (2)Chemical & Environmental Engineering, University of Arizona, Tucson, AZ

Oxidationof Biomolecules by Emerging Inorganic Nanoparticles

Antonia Luna-Velasco, Reyes Sierra-Alvarez, Jim A.Field

Departmentof Chemical and Environmental Engineering

Universityof Arizona, P.O. Box 210011, Tucson, Arizona 85721, USA

In recent years, inorganicnanoparticles (NPs) are finding increasingapplication in manufacturing and consumer products. The unusual properties ofthese emerging pollutants (e.g., very small size, high surface area andreactivity) has lead to growing public and regulatory concern about theirpotential toxicity. An important mechanism suggested for nanotoxicityis oxidative stress caused by reactive oxygen species (ROS). Testing thechemical generation of ROS by the interaction of NPs with biomoleculescould be a helpful tool for nanotoxicity screening. Inthis study, a rapid screening assay was developed to determine the chemicalgeneration of ROS by NPs through their interaction with biomoleculessusceptible to oxidation such as, L-3,4-dihydroxyphemylalanine(L-dopa) and 1,2-dihydroxybenzene (catechol). The assay utilized the dye dichlorofluorescin(DCFH), which is oxidized to its highly fluorescent product dichlorofluorescein(DCF) by ROS. Additionally, oxidation of proteins by NPs was evaluated using anELISA assay (based on detecting carbonyl groups in amino acid side chains).

Theinteraction of nanosized CeO_2 with L-dopa orcatechol significantly enhanced the oxidation of theROS indicator dye. Nanoscale CeO_2, $Fe_2O_3$and Fe^0 enhanced ROS production during the autooxidationof L-dopa by more than four-fold in reactions that were dependent on O_2. Mn_2O_3 NPs oxidized DCFHin the absence of O_2 or L-dopa, suggesting a direct oxidation of thedye. Mn_2O_3 NPs also significantly oxidized the protein bovineserum albumin indicating their potential reactivity with cell components. ROSgeneration by NPs (CeO_2, Mn_2O_3 and Fe^0)in aerobic aqueous suspensions was confirmed by electron paramagnetic resonance(EPR) signature for hydroxyl radicals with spin traps. The results taken as awhole indicate that NPs can generate ROS via chemical reactions with mediumcomponents and biomolecules susceptible to oxidation,such as L-dopa. The rapid assay with L-dopa and DCFH is a method proposed toscreen for chemical ROS production by NPs.

(292d) Demonstration of Nanoparticle-Bound Polymer Biodegradation and Resulting Nanoparticle Destabilization

Teresa L. Kirschling[1], Patricia L. Golas[2], Krzysztof Matyjaszewski[2], Kelvin B. Gregory[3], Gregory V. Lowry[3] and **Robert D. Tilton**[4], (1)Department of Chemical Engineering, Carnegie Mellon University, Pittsburgh, PA, (2)Department of Chemistry, Carnegie Mellon University, Pittsburgh, PA, (3) Department of Civil and Environmental Engineering, Carnegie Mellon University, Pittsburgh, PA, (4) Department of Chemical Engineering and Department of Biomedical Engineering, Carnegie Mellon University, Pittsburgh, PA

Many engineered nanoparticles are modified by adsorption or grafting of polymers onto their surfaces in order to impart colloidal stability in suspension, to promote adhesion with matrix materials, to increase biocompatibility, or to impart some particular biomolecular targeting capability. By controlling nanoparticle flocculation and surface deposition, surface-bound polymer layers strongly influence nanoparticle fate, transport, and subsequent biological impact in environmental systems. Biodegradation is one potential route to the breakdown of surface-bound polymers, but it is unknown whether such nanoparticle-bound polymers are bioavailable or if the surface confinement restricts microbial access to the chains so that they become recalcitrant to biodegradation. Most engineered nanoparticles are in a size range that cannot be internalized by bacteria, so the bound chains must be accessed extracellularly. Here we demonstrate, for the first time, that a polymer brush bound to nanomaterials can be bioavailable to microbes. Model core/shell nanomaterials with covalently attached polyethylene oxide (PEO) chains were prepared for this study. These were synthesized to obtain a nanomaterial in which biodegradation was the only available coating breakdown mechanism. Taking the form of a densely crosslinked bottlebrush copolymer with a hydrophobic core, these nanomaterials consisted of poly(ethylene oxide) methacrylate macromonomers crosslinked by hydrophobic divinyl benzene. PEO is a common constituent of consumer products, and PEO-based polymers are common, effective steric stabilizers for nanoparticles and other colloids. Accordingly, this an appropriate model system to represent a class of nanoparticle coatings that is likely to be introduced to the environment. The nanoparticles were approximately 30 nm in hydrodynamic radius and were colloidally stable in the culture media used in the biodegradation experiments. Microbial samples isolated from river water were enriched by supplying PEO homopolymer as the sole carbon source. The PEO-homopolymer degrading microbial enrichment cultures were transferred to fresh media, then supplied with PEO brush-bound nanoparticles as the sole carbon source. Biomass and CO_2 production measurements indicated that the particle-bound PEO brush was indeed bioavailable to these microbes. DNA sequencing indicated that the mixed culture responsible for degradation included *Novosphingobium sp.*, *Pseudomonas sp.*, and *Hydrogenophaga sp.* Brush biodegradation resulted in nanoparticle flocculation. Control experiments demonstrated that byproducts of microbial metabolism or cell debris did not cause the observed nanoparticle flocculation. This points to the need to understand biologically mediated transformations of nanoparticle coatings in order to understand the fate and transport of nanoparticles in the environment.

See more of this Session: Environmental Implications of Nanomaterials: Biological Interactions
See more of this Group/Topical: Topical C: Environmental Aspects, Applications, and Implications of Nanomaterials and Nanotechnology

Partitioning of Fullerene between Water and Synthetic Membrane Materials: Effects of Temperature and Membrane Compositions

Yeonjeong Ha, Howard M. Liljestrand, Lynn E. Katz

Department of Civil, Architectural and Environmental Engineering

The University of Texas at Austin, 1 University Station C1786, Austin, Texas, 78712

1. Introduction

As the usage of engineered nanomaterials increases, concerns on the unknown risks of these nanoparticles towards humans and the environment have grown. Despite these growing concerns about nanomaterials in aquatic environments, many unknowns remain. Investigating bio-uptake and bio-accumulation of nanomaterials is of critical importance not only to determine bioavailability but also to evaluate fate and transport in the environment of these compounds.

For hydrophobic contaminants and ionized organic chemicals, many models have been developed to determine their bio-concentration and toxicity. For example, octanol-water partitioning (K_{ow}) has been successfully used to evaluate the bioavailability of chemicals. More recently, liposome-water partitioning (K_{lipw}) has been proposed as being more appropriate than octanol-water partitioning (K_{ow}) because synthetic membrane vesicles are better surrogates and K_{lipw} can reduce thermodynamic differences from actual biota. For engineered nanomaterials, K_{ow} values of C_{60} (Jafvert et al., 2008) and carbonnanotubes (Petersen et al., 2010) have been determined. However, few attempts have been made to measure K_{lipw} of nanoparticles to investigate bioavailability of engineered nanomaterials.

Carbon fullerene (C_{60}), which was discovered in 1985 (Kroto et al., 1985), is a major material of nanoscale science and engineering. The production and application of fullerene have steadily increased due to increasing commercial interest in its unique properties, such as cage like structure, strong hydrophobicity, and high reactivity. It is estimated that more than 1500 tons per year of carbon nanoparticles including fullerenes and nanotubes are produced (Ringwood et al., 2009). However, there is limited information concerning the impact of fullerene on the natural environment and human health.

Bioavailability of fullerene in aquatic environment can be investigated by measuring partition coefficient between liposome and water (K_{lipw}) similar to that for other chemicals. Many researchers (Yamamoto et al., 2004, Kwon et al., 2006) have determined K_{lipw} using an equilibrium dialysis technique in which water and liposome solutions are separated by a dialysis membrane through which the chemical can diffuse but the liposome cannot. However this approach is less useful for fullerene, because the dialysis membrane can prevent the diffusion of engineered nanoparticles after their aggregation in water to a particle size much bigger than the molecular scale of hydrophobic or ionizing organic chemicals.

In this study, solid supported lipid membrane (SSLM) was chosen as a model biological membranes. Nonporous silica beads were used as the solids, and these were coated with the lipids. K_{lipw} of fullerene was investigated with various types of lipid membrane materials and temperatures. In addition, ternary lipid (liquid ordered phase lipid, liquid disordered phase lipid and cholesterol) mixture membrane vesicles were prepared at two different temperatures (above and below the transition temperature) to investigate whether K_{lipw} of nanoparticles can be influenced by the phase height difference between liquid disordered and liquid ordered/cholesterol. Lipids used in this study are 1,2 dimyristoyl-sn-glycero-3-phosphochloline (DMPC), 1,2 dioleoyl-sn-glycero-3 phosphocholine (DOPC), 1,2 dierucoyl-sn-glycero-3-phosphochloline (DEruPC), DOPC/ sphinogosylphosphochloline (SM) /cholesterol, DMPC/SM/cholesterol, DEruPC/SM/cholesterol.

2. Materials and Methods

2.1 Chemicals

Fullerene (C_{60}, 99.5+ %) was purchased from SES Research (Houston, TX). Unsaturated lipids such as 1,2 dimyristoyl-sn-glycero-3-phosphochloline (DMPC, C14:1, C14:1), dioleoyl-sn-glycero-3 phosphocholine (DOPC, C18:1, C18:1), and 1,2 dierucoyl-sn-glycero-3-phosphochloline (DEruPC, C22:1, C22:1), and sphinogosylphosphochloline (SM) were obtained from Avanti Polar Lipids. Cholesterol was purchased from Aldrich Chemical Co., and non-functionalized silica microspheres (mean diameter of 5.2 μm, 100 % solids content) were obtained from Bang's Laboratories.

2.2 Aqueous C_{60} solution

Fullerene has negligible solubility in water due to its strong hydrophobicity. However, many studies have shown that fullerene can form highly stable, negatively charged colloidal aggregates (nC_{60}) when it is exposed to water. Stable aggregated forms of fullerene have more than 8-11 orders of magnitude greater solubility than its hydrophobic form (Ruoff et al., 1993). Therefore, it is evident that the transition from its extreme hydrophobic form to water stable forms have critical implications for fullerene transport, transformation and biological effects (Isaacson et al., 2009).

The water stable suspensions of fullerene can be formed by various preparation methods in the laboratory. In this study, a modified SON/nC_{60} method was applied to make the aqueous C_{60} solution. Briefly, the fullerenes nanoparticles were formed by first adding pure fullerene to toluene. The solution was then mixed with a large volume of the deionized water and the toluene was removed by evaporation under ultrasonic treatment. The fullerene suspension in water was filtered through a 0.8 μm membrane filter. The mean size of C_{60} particles in water is in the range between 50 and 100 nm as measured by Dynamic Light Scattering (Model Zetasizer nano ZS).

2.3 Preparation of solid supported lipid bilayer

The synthetic membrane vesicles of each lipid compositions can be prepared using the thin film hydration technique followed by the rapid extrusion process. The detailed procedure is described elsewhere (Escher et al., 2000, Kwon et al., 2006). Briefly, the organic solution of lipid was evaporated using a nitrogen purging and the thin film residue was dissolved in water. After storage and equilibration at 4°C for a day, the lipid suspension was extruded several times through a polycarbonate membrane to reduce polydiversity of vesicles.

Solid supported lipid bilayers were formed using methods described in Bayerl et al., (1990) and Baksh et al., (2004). Silica beads were washed three times with methanol. The lipid suspensions were incubated at 30 °C for 30 min, and the lipid dispersions were sonicated (Branson 2000) for 15 min. The resulting vesicle dispersions were incubated at 60 °C for 5 min and then poured onto the dry glass beads with 60 sec rigorous vortex mixing. After vortexing, the silica beads and lipid dispersions mixture were centrifuged for 10 min at 2,000 g and then supernatants were discarded. Excess vesicles not adsorbed at the solid surface were removed by redispersing in water, vortexing, and the supernatants were discarded. The mass of lipids which were adsorbed on the silica beads was determined by measuring the difference between initial lipid concentration and supernatant concentration. Each lipid concentration was measured by TOC Analyzer (Tekmar Dohrmann Apollo 9000).

For solid supported ternary lipid mixture membrane vesicles, weight ratio of unsaturated lipids: SM: cholesterol was 2:1:1. Two replicate samples were prepared. One sample was stored at below the transition temperature (T_m), and the other one was first stored at above T_m overnight and then placed at below T_m to create phase height difference between liquid disordered lipid and liquid ordered lipid/cholesterol.

2.4 Determination of K_{lipw}

The solid supported lipid bilayer and aqueous fullerene dispersion were placed into 1.8 mL amber vial with PTFE/silicon septa. The reactors were incubated for 80 hr using a custom-made tumbling device (4 °C, 11 °C, and 25 °C) or shaker (30 °C and 50 °C, at 180 rpm). For ternary lipid mixture, vials were incubated at 25 °C with a tumbling device. The preliminary experiment showed that equilibrium was obtained after 80 hrs using both tumbler and shaker. After 80 hrs, each vial was placed for 1 hr to settle down the solid supported lipid bilayer. Then supernatants contained only free fullerene without lipid vesicles, and concentration of fullerene in supernatants (C_w) were measured using a Waters 2690 high-performance liquid chromatography system equipped with a Waters 996 photodiode array detector (Milford, MA, USA) after liquid extraction using toluene and 0.1 M of $Mg(ClO_4)_2$. Double or Triplicate samples were prepared for each test. The partition coefficient (K_{lipw}) was calculated using

$$K_{lipw}(L/kg-liposome)=\frac{C_{lip}}{C_w m}=\frac{C_0-C_w}{C_w m}$$

Where C_{lip} is the concentration of fullerene in the liposome side, which is same as the difference between the initial or control C_{60} concentration (C_0) and fullerene concentration in supernatants (C_w) and *m* is the concentration of liposome (kg-liposome/L).

3. Results and Discussion

Table 1 shows K_{lipw} values between water and DMPC ($K_{lipw,DMPC}$), and between water and DEruPC ($K_{lipw,DEruPC}$) at five different temperatures (4 °C, 11 °C, 25 °C, 30 °C, and 50 °C). The log K_{lipw} (L/kg) values range from 3.3 to 5.3. As far as we know, the K_{lipw} value of fullerene has not been reported. Some studies have concerned about Bioconcentration factor (BCF) of fullerene: Jafvert et al., (2008) estimated log BCF values of fullerene as 4.51, 5.37, 5.71 for cod, earthworms and salmon, respectively, and Tervonen et al., (2010) measured log BCF values of fullerene for *Daphnia magna* ranged from 3.30 and 3.88. Those reported values are generally consistent with the results of this study.

Table 1. Liposome – Water Partition Coefficients (K_{lipw}, L/kg-liposome) of fullerene obtained at five different temperatures using 1,2 dimyristoyl-sn-glycero-3-phosphochloline (DMPC) and 1,2 dierucoyl-sn-glycero-3-phosphochloline (DEruPC)*

	4 °C	11 °C	25 °C	30 °C	50 °C
$K_{lipw,DMPC}$	1.96×10^3	4.34×10^3	7.87×10^3	1.52×10^4	2.18×10^4
$K_{lipw,DEruPC}$	9.92×10^3	1.07×10^4	2.64×10^4	5.62×10^4	1.89×10^5

*Average values of duplicates analysis are used

Both $K_{lipw,DOPC}$ and $K_{lipw,DEruPC}$ values of fullerene increase with increasing temperatures (Figure 1). It has been reported that for most of hydrophobic chemicals such as EDCs (Kwon et al., 2007), K_{lipw} values with liquid crystalline phase liposomes decreased with increasing temperature. Therefore, partitioning processes of fullerene between water and liposome can be different from those of hydrophobic chemicals.

At all different temperatures, log $K_{lipw,DEruPC}$ values are higher than log $K_{lipw,DMPC}$ by more than 0.4 log units (Figure 1). Based on this result, it can be concluded that K_{lipw} values of fullerenes are larger for lipid bilayers composed of longer acyl chains. Wong-Ekkabut et al., (2008) indicated that when aggregated fullerenes are located close to the lipid head group, only small clusters form and rapidly penetrate into the bilayer and then disaggregate after entering the membrane. This result shows that, despite of large size of fullerene aggregates, which are from ten to a few hundreds of nanometers, they can penetrate through lipid bilayers. Therefore, the length of acyl chain of lipids can affect the K_{lipw} of fullerenes. However, more work is required to confirm this hypothesis.

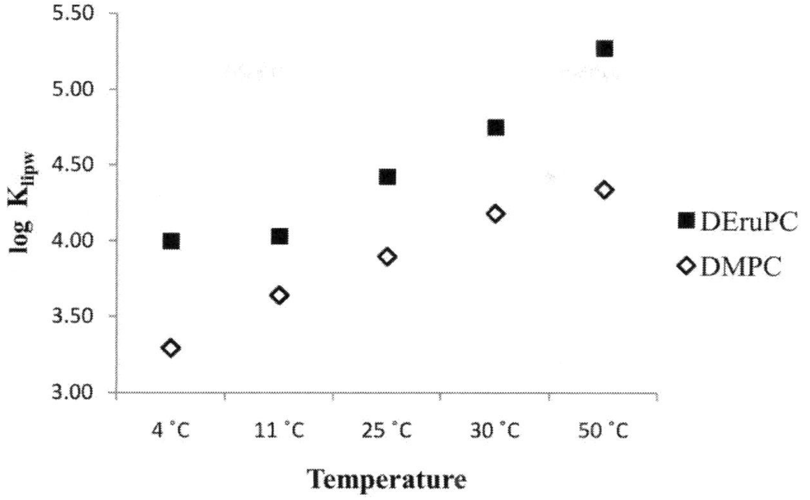

Figure 1. Effects of temperatures on lipid – water partitioning coefficients (K_{lipw}, L/kg-liposome) of fullerene using DMPC and DEruPC (average values of duplicates analysis are used).

4. References

1. Jafvert, C.T., Kulkarni, P.P. (2008), "Buckminsterfullerene's (C_{60}) octanol-water partition coefficient (Kow) and aqueous solubility," *Environ. Sci. Technol.* 42. pp. 5945-5950.
2. Petersen, E.J., Huang, Q., Weber, W.J.(2010), "Relevance of octanol-water distribution measurements to the potential ecological uptake of multi-walled carbon nanotubes," *Environ. Toxicol. Chem.* 29. pp. 1106-1112.
3. Kroto, H.W.,Heath, J.R., O'Brien, S.C., Curl, R.F., Smalley, R. E.(1985), "C60-Buckminsterfullerene," *Nature.* 318. pp. 162-163.
4. Ringwood, A.H., Levi-Polyachenko,N., Carroll, D.L. (2009), " Fullerene Exposures with Oysters: Embryonic, Adult, and Cellular Responses," *Environ. Sci. Technol.* 43. pp. 7136-7141.
5. Yamamoto, H., Liljestrand, H.M., Shimizu, Y. (2004), "Effects of dissolved organic matter surrogates on the partitioning of 17ß-Estradiol and p-Nonylphenol between synthetic membrane vesicles and water," *Environ. Sci. Technol.* 38(8). 2351-2358.
6. Kwon, JH., Liljestrand, H.M., Katz, L.E. (2006), "Partitioning of moderately hydrophobic endocrine discruptors between water and synthetic membrane vesicles," *Environ. Toxicol. Chem.* 25(8). pp.1984-1992.
7. Ruoff, R. S., Tse, D. S., Malhotra,R., Lorents, D.C. (1993), "Solubillity of C_{60} in a variety of solvents," *J. Phys. Chem.* 97. pp. 3379-3383.
8. Isaacson, C.W., Kleber, M., Field, J.A. (2009), "Quantitative analysis of fullerene nanomaterials in environmental systems: A critical review," *Environ. Sci. Technol.* 43. pp. 6463-6474.

9. Escher, B.I., Schwarzenbach, R.P. (2000), "Evaluation of liposome – Water partitioning of organic acids and bases. 1. Development of a sorption model," *Environ. Sci. Technol.* 34(18). pp.3954-3961.
10. Bayerl, T.M., Bloom, M. (1990), "Physical properties of single phospholipid bilayers adsorbed to micro glass beads," *Biophys. J.* 58. pp. 357-362.
11. Baksh, M.M., Jaros, M., Groves, J.T. (2004), "Detection of molecular interactions at membrane surfaces through colloid phase transitions," *Nature.* 427. pp. 139-141.
12. Tervonen, K., Waissi, G., Petersen, E.J., akkanen, J., Kukkonen, J.V. (2010), "Analysis of fullerene-C60 and kinetic measurements for its accumulation and depuration in Daphnia magna," *Environ. Toxicol. Chem.* 29. pp. 1072-1078.
13. Kwon, JH., Liljestrand, H.M., Katz, L.E., Hiroshi, Y. (2007), "Partitioning thermodynamics of selected endocrine disruptors between water and synthetic membrane vesicles: Effect of membrane compositions", *Environ. Sci. Technol.* 41. pp. 4011-4018.
14. Wong-ekkabut, J., Baoukina, S., Triampo, W., Tang, I.M., Tieleman, D.P., Monticelli, L. (2008), "Computer simulation study of fullerene translocation through lipid membrane", *Nature Nanotechnology*. 3. pp. 363-368.

(292f) Single Wall Carbon Nanotubes Enter Cells by Endocytosis and Not Membrane Penetration

Peter N. Yaron[1], Brian D. Holt[2], Phillip A. Short[2], Mathias Lsche[3], Mohammad F. Islam[4] and Kris N. Dahl[5], (1)Department of Chemical Engineering, Carnegie Mellon University, Pittsburgh, PA, (2) Biomedical Engineering, Carnegie Mellon University, Pittsburgh, PA, (3)Physics, Carnegie Mellon University, Pittsburgh, PA, (4)Material Science, Carnegie Mellon University, Pittsburgh, PA, (5) Chemical Engineering, Carnegie Mellon University, Pittsburgh, PA

Carbon nanotubes are increasingly being tested for use in cellular applications. Determining the mode of entry is essential to control and regulate specific interactions with cells, to understand toxicological effects of nanotubes, and to develop nanotube-based cellular technologies. We investigated cellular uptake of Pluronic copolymer–stabilized, purified ~145 nm long single wall carbon nanotubes (SWCNTs) through a series of complementary cellular, cell-mimetic, and *in vitro* model membrane experiments. SWCNTs localized within fluorescently labeled endosomes as measured by fluorescence lifetime imaging microscopy. Confocal Raman spectroscopy showed a dramatic reduction in SWCNT uptake into cells at 4 °C compared with 37 °C. Electrochemical impedance spectroscopy and Langmuir monolayer film balance measurements showed that Pluronic-stabilized SWCNTs associated with membranes but did not possess sufficient insertion energy to penetrate through the membrane. SWCNTs associated with vesicles made from plasma membranes but did not rupture the vesicles. These measurements, combined, demonstrated that Pluronic-stabilized SWCNTs did not penetrate the membrane via non-specific physical interactions but entered cells via endocytosis, and association of SWCNTs to membrane likely increased uptake. These results suggest that specific targeting of SWCNTs for use in therapeutics should focus on endocytosis as the preferential method for the cellular uptake.

See more of this Session: Environmental Implications of Nanomaterials: Biological Interactions
See more of this Group/Topical: Topical C: Environmental Aspects, Applications, and Implications of Nanomaterials and Nanotechnology

(369a) Modeling Aggregation and Size Distribution of Nanoparticles In Aqueous Suspensions

Haoyang H. Liu, Chemical and Biomolecular Engineering, University of California, Los Angeles, Los Angeles, CA, Sirikarn Surawanvijit, Chemical&Biomolecular Engineering, University of California, Los Angeles, Los Angeles, CA, G. Orkoulas, Department of Chemical and Biomolecular Engineering, University of California, Los Angeles, Los Angeles, CA and Yoram Cohen, Chemical & Biomolecular Engineering, University of California, Los Angeles, Los Angeles, CA

The size distribution of the engineered nanoparticles (eNMs) and their aggregates in aqueous suspensions is crucial for predictions of the fate and transport of nanoparticles in the environment. Diffusion, sedimentation and interaction of nanoparticles with solid surfaces (including sedimentation), as well as living organisms are all affected by the size distribution of eNMs. In order to obtain the nanoparticle size distribution over the expected ranges of water chemistry and nanoparticle properties, predictive models are needed to assess the aggregation tendency of eNMs. In the present work, a constant-number kinetic Monte Carlo based model was developed to predict the aggregation rate and size distribution of nanoparticles. The modeling approach followed the "particles in a box" simulation method while allowing for particle sedimentation. The likelihood of nanoparticle aggregation was determined, based on the classical DLVO theory, via calculation of the aggregation probability of nanoparticle pairs. In developing the model, the reasonable minimum number of simulation particles and simulations instances were first evaluated to ensure convergence of the stationary statistics, demonstrating the ability to use parallel computing to dramatically decreasing simulation time. Model performance with respect to the predicted size distribution of eNMs was evaluated via comparison with experimental dynamic light scattering (DLS) measurements and with available literature data under varying conditions (e.g., pH, ionic strength, size distribution of the primary nanoparticles and nanoparticle surface charge). Model performance compared favorably with experimental DLS data demonstrating the potential of the present modeling approach for environmental analysis of the transport and fate of nanoparticles

See more of this Session: Environmental Implications of Nanomaterials: Fate and Transport
See more of this Group/Topical: Topical C: Environmental Aspects, Applications, and Implications of Nanomaterials and Nanotechnology

(369b) Stability of Engineered Nanoparticles In Aqueous Systems: Elucidating the Roles of Capping Agents and Natural Organic Matter

Jeffrey A. Nason, School of Chemical, Biological and Environmental Engineering, Oregon State University, Corvallis, OR

The rapid development of novel nanomaterials and their incorporation into consumer products has not been paralleled with an equal effort investigating the possible environmental implications. Increasingly, the properties of engineered nanoparticles (ENPs) are tailored for specific applications through the use of organic capping agents. Furthermore, when released into the environment it is probably that ENPs will interact with natural organic matter (NOM). The objective of this work is to develop an improved understanding of the roles of synthetic capping agents and NOM in controlling the aggregation of engineered nanoparticles (ENPs) in aquatic environmental systems. Using a suite of ligand-stabilized gold nanoparticles (AuNPs) and various NOM isolates, the specific aims of the project are to: (1) Correlate the physicochemical properties of capping agents with interactions between capped ENPs in aquatic systems; and (2) Correlate the physicochemical properties of capping agents and NOM with interactions between capped ENPs and NOM in aquatic systems. ENP stability with respect to aggregation is being assessed using the zeta potential and aggregation rates measured using time-resolved dynamic light scattering. Results indicate that NOM is adsorbed onto the AuNP surfaces, alterning surface charge and providing stability with respect to aggregation. Different classes of NOM stabilize AuNPs to different extents. The specific mechanisms of NOM-AuNP interactions are being probed using advanced surface and solution phases spectroscopic techniques (e.g., fluorescence, Raman, and EDX). By using a group of well-characterized ENPs that vary only with respect to their surface functionality, we aim to correlate environmental behavior with ENP properties in structure-activity type relationships. For example, results indicate that the stabilizing effects of NOM are related to the molecular weight of the specific NOM fractions. This work will provide an improved ability to predict environmental fate based on NP characteristics and yield the feedback necessary for the design of safer nanomaterials.

See more of this Session: Environmental Implications of Nanomaterials: Fate and Transport
See more of this Group/Topical: Topical C: Environmental Aspects, Applications, and Implications of Nanomaterials and Nanotechnology

(369c) Removal of Nanoparticles In Semiconductor Manufacturing Effluents Using Porous Media Filtration

Jeff Rottman, Reyes Sierra and Farhang Shadman, Chemical and Environmental Engineering, University of Arizona, Tucson, AZ

The use of engineered nanoparticles in industrial manufacturing has rapidly increased due to their unique electronic, optical, thermal and photoactive properties. Nanoparticles are used in various fields including textiles, biomedical applications, food and agriculture and semiconductor manufacturing. Frequently used nanoparticles include titania (TiO_2), which is commonly used in pigments and as a photocatalyst, and silica (SiO_2), which is commonly used as an additive in chemical mechanical planarization (CMP) slurries in semiconductor manufacturing. Along with the benefits these nanoparticles provide, numerous concerns into the environmental, safety, and health effects of the released particles arise. Understanding the fundamental interactions of these nanoparticles with typical treatment technologies will allow for improved design of abatement technologies. Porous media filtration is a common technique used to remove particulates from water. The simplicity and ubiquity of this process makes it ideal for the removal of nanoparticles in waste streams. The principles of nanoparticle transport in porous media also relate to the environmental fate of these nanoparticles in soil and ground water.

Custom nanoparticles of silica have been developed which have a fluorescent core and controllable final size. These particles, due to their fluorescence, can be measured at significantly lower concentrations than in current studies. Additionally, the ability to control the final size of the particles provides a true comparison of size effects on nanoparticle retention. A test bed, equipped with real-time and on-line spectrometry, was developed and used to study the dynamics of capture and removal of both TiO_2 and these SiO_2 nanoparticles from wastewater. Studies were conducted comparing the effect of key solution, nanoparticle, and bed characteristics, including pH, ionic strength, the presence of surfactants, heavy metals, and dispersants, as well as particle and collector sizes. A process model is under development for analysis of data from this novel experimental method and for elucidating the fundamental parameters affecting nanoparticle retention and removal in such systems.

See more of this Session: Environmental Implications of Nanomaterials: Fate and Transport
See more of this Group/Topical: Topical C: Environmental Aspects, Applications, and Implications of Nanomaterials and Nanotechnology

(369d) Adsorption of Gold Nanoparticles and Humic Acid On Activated Carbon Used In Drinking Water Treatment

Holly A. Stretz, Department of Chemical Engineering, Tennessee Technological University, Cookeville, TN, Vasanta Pallem, Department of Chemical Engineering and Center for the Management, Utilization and Protection of Water Resources, Tennessee Technological University, Cookeville, TN and Martha J. M. Wells, Department of Pharmaceutical Sciences, Midway College School of Pharmacy, Paintsville, KY

Granular activated carbon (GAC) is primarily used in drinking water treatment processes for removing dissolved organic matter comprising of humic and fulvic acids (DOM), and other pollutants. The increasing applications of gold nanoparticles (GNPs) in diagnostic imaging, chemical and biological sensing, also present potential concerns for adsorption and removal of GNPs in water treatment units, following post-clinical discharge. The current study employs spectroscopic separation to evaluate the extent of adsorption of citrate coated GNPs with core size 50 nm (GNPs-50), and HA separately and in mixtures. Modified Freundlich isotherms were used to describe the adsorption behavior of humic acid and gold in mixtures. The novelty of the approach was to examine the adsorption behavior for synergistic or antagonistic effects when all three materials (gold nanoparticles, HA and activated carbon) were present, and to carefully account for the fluorescence of the controls in determining fractional fluorescence prior to establishing the isotherms. The GNPs-50 were completely adsorbed on activated carbon when present individually, however in the presence of HA the adsorption of GNPs-50 decreased significantly. The adsorption of humic acid also decreased in the presence of GNPs-50 showing an antagonistic effect. Therefore, the mixtures of HA/GNPs-50 present significant implications for the fate and removal of GNPs and HA in water processing units.

See more of this Session: Environmental Implications of Nanomaterials: Fate and Transport
See more of this Group/Topical: Topical C: Environmental Aspects, Applications, and Implications of Nanomaterials and Nanotechnology

(369e) High Throughput Collection and Detection of Environmental Nanoparticles

Fanxu Meng[1], Serdar Ozturk[1], Maria D. King[2], Yassin A. Hassan[3] and Victor M. Ugaz[1], (1)Artie McFerrin Department of Chemical Engineering, Texas A&M University, College Station, TX, (2) Department of Mechanical Engineering, Texas A&M University, College Station, TX, (3)Departments of Nuclear and Mechanical Engineering, Texas A&M University, College Station, TX

The increasing prevalence of manufactured nanoparticles in the environment poses new risks to human health, pointing to the need for advanced nanoparticle sampling and analysis approaches capable of continuously monitoring room-sized volumes so that dynamic changes in exposure levels can be rapidly detected. Here we describe our efforts to address this need through development of a new technique that enables nanoparticles to be collected and concentrated at very high flow rates, so that they can be directly detected using a continuous-flow microfluidic approach coupled with particle sizing analysis. Nanoparticle detection is accomplished by harnessing interfacial fluorescence phenomena that uniquely emerge in a microfluidic environment. Fundamental parameters including sensitivity and specificity are characterized to establish detection limits for comparison with current-generation instrumentation. We then explore feasibility of coupling this component with an autonomous Wetted Wall Cyclone (WWC) collector developed in our laboratory and fabricated at Northrop Grumman Inc. This system is capable of performing continuous air sampling at 400 L/min with a cutpoint of 500 nm. Performance was characterized by sampling colloidal (anatase, 5-30 nm) and amorphous powder (5-30 nm) TiO_2 nanoparticles aerosolized and collected in a laboratory flowcell, and comparing the results against a standard glass fiber filter sampler. Environmental monitoring capability was assessed by sampling air in a freshly painted classroom for 24 hours at 100 L/min. Each hydrosol sample was weighed and quantitated alongside suspensions extracted from the reference filter using inductively coupled plasma mass spectrometry. This analysis indicated that the WWC system achieved collection efficiencies of 83.5% and 56.5% for aerosolized colloidal and amorphous power TiO_2 nanoparticles, respectively. The level of TiO_2 in the painted room air was around 0.2 ppt, however, SiO_2 was also detected at 1 ppb. Spectral analysis of the interfacial fluorescence offers potential to enable chemical fingerprinting to be rapidly performed in order to obtain a detailed picture of nanoparticle exposure levels while simultaneously providing the ability to respond to sudden environmental changes

See more of this Session: Environmental Implications of Nanomaterials: Fate and Transport
See more of this Group/Topical: Topical C: Environmental Aspects, Applications, and Implications of Nanomaterials and Nanotechnology

NANOMATERIALS: THE LATEST EMERGING ENVIRONMENTAL CONTAMINANT IDENTIFICATION AND MEASUREMENT

Presented by:
Katrina E. Varner
US Environmental Protection Agency/Office of Research & Development
Las Vegas, NV, USA

BIOGRAPHY:

My primary research focus as an environmental research scientist is method development and evaluation of parameters monitoring for contaminants and hazardous waste. Method and monitoring development involve the detection, speciation, and identification of nanomaterials in various media. This includes the development of screening techniques for groundwater and soil. My goal is to advance analytical techniques for the characterization and remediation of nanomaterials in various media including water, soil, and tissue to estimate the daily exposure of animals and humans to these materials.

Introduction

The basic research presented is a novel tool for the detection, characterization, quantification, and monitoring of nanomaterials in the environment. EPA is becoming more proactive concerning environmental issues. Emerging contaminants is an area EPA is investigating to determine if new products introduced by man are harmful to people or the environment. Nanomaterials are the latest emerging chemical group. These materials are being incorporated into commercial products at a faster rate than the development of knowledge and regulations to mitigate potential environmental impacts. Product technology is rapidly advancing, but are we, scientists and engineers, keeping pace with understanding and communicating information about potential health risks and effect to the public from these products? EPA's Office of Research and Development charged its laboratories to perform research on nanomaterials to gain understanding of the transformation, transportation, and longevity of these materials. The National Exposure Research Laboratory is specifically responsible for conducting nanotechnology studies.

EPA's National Exposure Research Laboratory-Environmental Sciences Division is researching the detection, characterization, quantification, and monitoring of nanomaterials in the environment. Before the possible risks of these materials in the environment can be determined, methods are needed for their collection, separation, detection, identification and quantification. The areas of research encompass the sources, fate and transport, exposure, and mitigation or prevention of the risks associated with nanomaterials. This research requires technology be applied to detect, identify, quantify and minimize exposure to engineered nanomaterials in the environment. Currently, nano-scale metallic and carbon based particles are not detectable in water or soil.

EPA is concerned with the transport of nanomaterials in the environment which requires understanding transfer mechanisms within environmental media, and measurement of various environmental parameters and nanomaterials parameters. Nanomaterial parameters include particle size and charge, chemical or elemental composition and surface modifications. Environmental media parameters include pH, ionic strength, flow rate, composition, and the presence of naturally occurring (such as dissolved organic carbon) as well as anthropogenic contaminant. Understanding interactions between nanomaterials and environmental media is also critically important.

The goals of this research are to develop and test novel monitoring and characterization methods for engineered nanomaterials (ENMs) with the initial focus on carbon nanoparticles (e.g. fullerenes) and metal nanoparticles (silver and gold) using electrochemical analyses. Increasing ENM content in consumer products may have a negative environmental impact. Many nano-enabled products including fullerene lotions and cosmetics contain reactive oxygen species that are readily introduced as non-point source pollution into the environment. Application onto the skin might pose more problems that may include immediate health concerns. Measurement of these materials is challenging because of their low concentrations and the presence of natural and anthropogenic nanomaterials containing the same elements. We are investigating the use of electrochemical probes with Quartz Crystal Microbalance (QCM) to quantitatively analyze for C_{60} versus C_{70}. The initial environmental matrix of water was utilized in the laboratory and further environmental matrices analyzed will include ground and surface water, soil and sediment.

Approach

A QCM gold electrode is modified with gold nanoparticles (AuNPs) prior to fictionalization using N-dicyclohexylcarbodiimide/N-hydroxysulfosuccinimide (DCC/NHS) chemistry (Figure 1). Decrease in frequency or mass increase is used to measure the adsorption/enumeration of C_{60} by the beta-cyclodextrin amine (ß–CD-NH_2) cavity attached on the QCM.

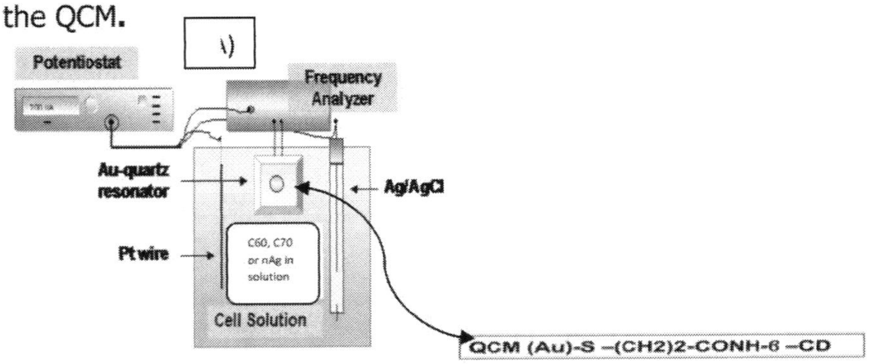

(B)

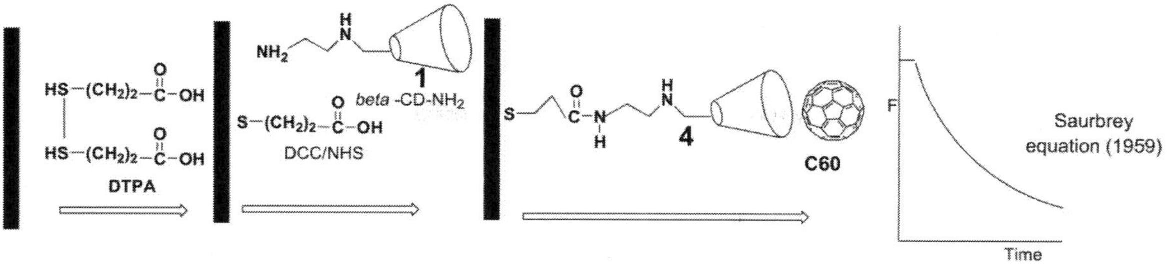

Δf is the change in frequency to coating (Hz), f0 is the frequency of quartz plate (MHz), Δm is the mass of engineered nanomaterial (in pico-grams), A is the area of quartz plate or electrode

Figure 1: *Schematic of modified QCM (A); Sensor design and surface modification using N-DCC/NHS chemistry (B).*

Discussion

The QCM is used to measure mass changes as well as the in-situ measurements of charge, current, potential and the resistance changes following the uptake of the particles. The unique sugar complex bucket-like structure is used to capture the spherical soccer ball-like structured C_{60}. Using amino modified ß–CD (a water-soluble polymer with synthetic receptor), the interactions of selected nanostructured materials (e.g. gold, silver, and fullerenes) are directly monitored by measuring the time-dependent changes in series resonant frequency by the QCM sensor. QCM provides a relationship between the resonance oscillation frequency of piezoelectric crystal and the interfacial mass change. The data are used to determine the quantity of the nanomaterial and comparisons are made with conventional methods such as Ultra Visible Spectroscopy (UV-Vis), Transmission Electron Microscopy (TEM) and Field Emission Scanning Electron Microscopy (FESEM). When the QCM is saturated with the nanomaterial, the attached nanomaterials are quantitatively released from the probe during a regeneration step washing with acid or alkaline hydrogen peroxide due to its selective association and dissociation behavior with C_{60}. And measure the concentration levels of C_{60} in water using simple QCM mass changes Application of a voltage between electrodes results in a shear deformation of the crystal

Results

Reproducible results are obtained using AuNPs in the immobilization chemistry. Figure 2A shows a plot of frequency versus time for the control, the QCM (AuNPs)-S–$(CH_2)_2$-CONH-ß–CD-NH_2 electrode in toluene. The figure demonstrates that no adsorption was noted on the modified QCM. In contrast, the frequency decreased (evidence of mass increase according to the Saurbrey equation) in the presence of 0.14 mg of C_{60} ml toluene (Figure 2B). These results suggest that the sensor can detect C_{60} in solution.

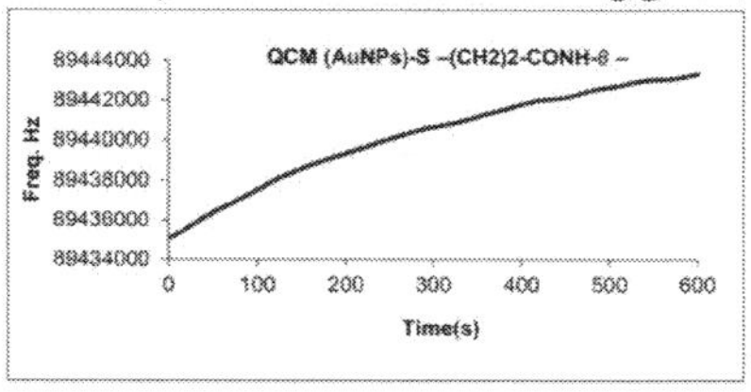

(A)

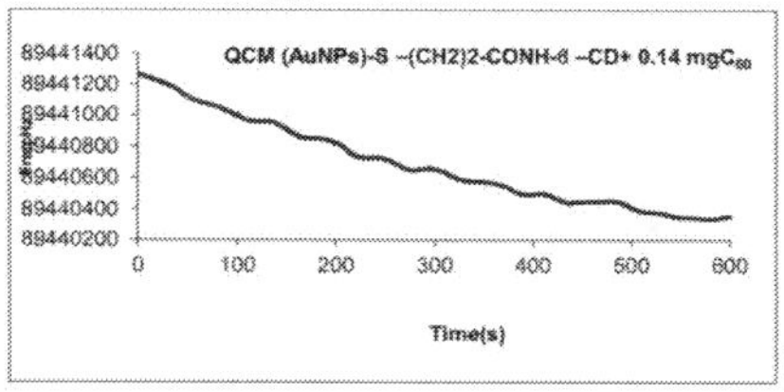

(B)

__Figure 2:__ Frequency vs. time for QCM (Au)-S–$(CH_2)_2$-CONH-β–CD in presence of (A) toluene only (B) 0.14 mg C 60 + toluene (run time = 600 seconds)

Figure 2B shows that the decrease in frequency continued up to 600 seconds. Additional frequency vs time change data showed that at 700s the maximum adsorption of C_{60} from 0.14 mg C_{60} /ml toluene was reached. This is observation indicated that all available active sites on the sensor were saturated. Sensor calibration data also showed that at much lower concentrations, the sensor can be calibrated up to 700 seconds. The QCM sensor has been tested for varying C_{60} concentrations.

Characterization

SEM images show the expected gold SEM micrograph. SEM micrographs of a clean modified QCM electrode (Figure 3A) and after exposure to C_{60} in toluene are shown. In the presence of C_{60} the SEM micrograph showed evidence of near spherical nanoparticles attributed to C_{60} clusters (Figure 3B), which were not observed on the pre-cleaned bare QCM.[1, 2] The presence of fullerene clusters of about 200 nm on the sensor surface suggests the aggregation of fullerenes since they exist as fractal systems which is consistent with previous studies.[3]

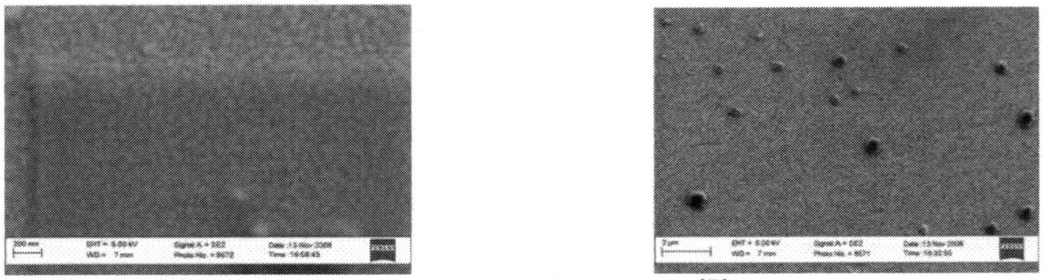

(A) (B)

__Figure 3:__ Pre-cleaned QCM, image @ 200nm (A) and QCM after evaporating drop of 2.8 mg C_{60}/toluene solution, image@ 200nm (B)

Conclusions

The outcome of this research is to provide quantitative data to exposure risk assessors that make possible predicative estimates of compounds most likely to be of concern for the protection of human health and the environment. The ultimate goal, to discover and define previously unknown or poorly understood vulnerabilities of nanomaterials, is being accomplished. The successful demonstration using the synthesis of β–CD-NH_2, its attachment and modification of QCM transducer via DCC/NHS chemistry is a promising new tool to address that goal. The QCM sensor showed specificity for C_{60} extraction when compared to the control solution by measuring the mass changes as well as the in-situ measurements of charge, current, potential and the resistance changes following the absorption and isolation of the particles to determine the presence of nanomaterials. A new practical measurement method for engineered, C_{60}, fullerenes is demonstrated as a new analytical tool; the first nanosensor to measure ENMs utilizing a size exclusive approach for the quantitative measurement of fullerenes.[4]

References:

1. Sadik, O.: Kikandi, S.; Wang, Q.; Varner, K. "Sensors as Tools for Quantitation, Nanotoxicity and Nanomonitoring Assessment of Engineered Nanaomaterials", *J. Environ. Monit.* **2009**, 11, 1782-1800.

2. Liu, Y.; Yang, W.; Chen, Y. Chemical Communications, **2005**, 4208-4210

3. Kikandi, S; Okello, V.; Wang, Q.; Sadik, O.; Varner, K.; Burns, S. "Size-Exclusive Nanosensor for Quantitative Analysis of Fullerene C_{60}", *Environ Science and Tech.* Accepted **2011**.

4. Bulavin, L.; Adamenko, I.; Yashchuk, V.; Ogul'chansky, T.; Prylutskyy, Y.; Durov, S.; Scharff, P. "Self-Organization C60 Nanoparticles in Toluene Solution", J. *of Molecular Liquids* **2001**, 93, (1-3), 187-191.

Work performed by:

Samuel Kikandi (Student Services Contract #RFQ-RT-08-00130), Qiong Wang, Veronic Okello under guidance and mentorship of Omowunmi Sadik, Director, Center for Advanced Sensors & Environmental System, State University of New York-Binghamton, Department of Chemistry, POB 6000, Binghamton, NY 13902-6000

(462b) Enabling In Situ Real-Time Characterization of Interfaces with Quartz Crystal Microbalance with Dissipation Monitoring

Mark A. Poggi, Q-Sense and Farfield, Biolin Scientific, Linthicum Heights, MD, Archana Jaiswal, Q-Sense, Biolin Scientific, Linthicum Heights, MD and Matthew Dixon, Biolin Scientific, Linthicum Heights, MD

There is a growing need for new technologies to quantitatively measure the surface properties of biological and material interfaces. One technique in particular, the Quartz Crystal Microbalance with Dissipation Monitoring (QCM-D), fulfills the need for monitoring real-time dynamic adsorption and desorption phenomena. Capable of operating in liquids and in real-time, QCM-D provides a powerful approach to analyze the *in situ* thickness, structural, and viscoelastic properties of molecules and materials at interfaces. This presentation will focus on the application of QCM-D to monitoring self assembly processes. Examples to be discussed will be recent research done in the areas of: protein deposition or fouling, polyelectrolyte multilayer build-up, nanoparticle/nanotube assembly, lipid membrane fuel cell interfaces. The QCM-D technique has allowed scientists to achieve a more fundamental understanding of how dynamic materials behave on the molecular scale and current research in several of these areas will be highlighted.

See more of this Session: Life Cycle Studies Associated with Nanomaterials and Nano-Enhanced Products
See more of this Group/Topical: Topical C: Environmental Aspects, Applications, and Implications of Nanomaterials and Nanotechnology

(462c) Mobility and Deposition of Silver Nanoparticles On Silica Surfaces Under Environmentally Relevant Conditions

B. Reginald Thio, Engineering Product and Design, Singapore University of Technology and Design, Singapore, Singapore, **Milka O. Montes**, Bren School of Environ. Sci & Mgmt, University of California, Santa Barbara, Santa Barbara, CA, Mahmoud A. Mahmoud, School of Chemistry and Biochemistry, Georgia Institute of Technology, Atlanta, GA and Arturo A. Keller, Bren School Environ. Sci & Mgmt, University of California, Santa Barbara, Santa Barbara, CA

Increasing use of nanomaterials in commercial products has led to widespread concerns about their potential environmental and health impacts. To better understand the transport, fate and behavior of nanoparticles in aquatic systems, it is essential to understand their interactions with different components of natural waters including natural organic matter (NOM) over a broad range of physicochemical conditions. Ag nanoparticles are widely used in many consumer applications such as antibacterial socks, and pigments. Deposition and mobility rates and the effects of NOM on particle deposition, as well as transport behavior of Ag nanoparticles onto a silica surface were evaluated using thin layer chromatography (TLC), and quartz crystal microbalance with dissipation (QCM-D) over a broad range of solution conditions.

See more of this Session: Life Cycle Studies Associated with Nanomaterials and Nano-Enhanced Products
See more of this Group/Topical: Topical C: Environmental Aspects, Applications, and Implications of Nanomaterials and Nanotechnology

(462d) Effect of the Aggregation of TiO2 Nanoparticles On Their Fate and Transport In Natural Waters

Mara Marta Fidalgo[1], Marina B. Romanello[1], Liliana M. Bertini[1], Linna Du[2] and Jianhong Ren[2], (1) Department of Chemical Engineering, Instituto Tecnolgico de Buenos Aires, Buenos Aires, Argentina, (2)Department of Environmental Engineering, Texas A&M University Kingsville, Kingsville, TX

Nanotechnology is a growing area with many promising and revolutionary advances in a variety of fields, from medicine to sporting good design and personal care products. One of the most widely applied nanomaterial is nanoTiO_2. Although it is a well-known material at the macroscale, the environmental implications of the special properties exhibited by TiO_2 in its nanoform are not fully understood. Of particular interest are the aggregation processes in natural waters, the interaction with natural organic matter, and the effect of the presence of divalent ion such as calcium and magnesium. The different conditions of the water matrix that TiO_2 nanoparticles may encounter in surface and groundwater will influence their physico-chemical properties, and thus modify their surface charge, aggregation, and attachment characteristics. These properties are key to determine the fate and transport of nanomaterials in the environment, and therefore a complete understanding of the transformations that TiO_2 undergo in a natural setting is necessary to assess the risk posed to the environment by products containing this nanomaterial.

The size and surface charge of TiO_2 nanoparticles was investigated as a function of pH for a variety of relevant conditions, including different levels of ionic strength, concentration of natural organic matter and non-indiferent cations by Dynamic Light Scattering (DLS) using a Nanosizer Z90 (Malvern, UK) and by means of settling column experiments for those cases when DLS was not applicable. By size the authors refer to the measured equivalent hydrodynamic diameter of the aggregates formed by the nanoparticles. DLVO theory was applied to model the experimental conditions explored, and generally good agreement was observed when aggregate size is taken as effective particle size for the calculations.

Aggregate size and morphology were investigated by Scanning Electron Microscopy using a FESEM (JEOL JSM-6701F). To preserve the aggregates conditions as close as possible to their original form in suspension, a substrate was placed vertically in each suspension and the sample to be imaged was formed by aggregates that attached during the descent of the liquid level as the suspension evaporates. Special care was taken so that the change in concentration of the constituents in the solutions did not vary significantly during this process, taking into account the ratio of the relative change of volume to the initial volume of sample suspension and interrupting the process accordingly. Prior to imaging, the samples were gold coated with a sputter coater (Denton Vacuum Desk IV) to provide a conductive surface.

Sand column transport experiments were conducted for some characteristic conditions to investigate the mobility of nanoTiO_2 in porous media. The results were modeled applying filtration theory.

It was observed that aggregation is dependent on pH, shear (mixing conditions), presence of natural organic matter and divalent ions through their effect in the zeta potential of the particles although this parameter alone was insufficient to explain the different size of the aggregates. Attachment to the sand grains during flow in the porous media followed a similar trend.

This work provides critical insights on the physico-chemical properties of TiO_2 nanoparticles suspended in environmentally relevant aqueous matrixes. The information presented is key to understand the interaction with the natural environment: their fate and transport on groundwater and surface water, their bioavailability and their subsequent effect on the biota.

See more of this Session: Life Cycle Studies Associated with Nanomaterials and Nano-Enhanced Products
See more of this Group/Topical: Topical C: Environmental Aspects, Applications, and Implications of Nanomaterials and Nanotechnology

(462e) Application of Screening-Level Life Cycle Assessment to Emerging Nanoproducts: Nanosilver Textiles and CNT Electronics

David E. Meyer, National Risk Management Research Laboratory, U. S. Environmental Protection Agency, Cincinnati, OH and Venkata K. K. Upadhyayula, Oak Ridge Institute of Science and Education (ORISE), Oak Ridge, TN

As global concerns for environmental sustainability grow, emerging technologies must be scrutinized early in deployment to insure they represent the most environmentally conscious option. Screening-level Life Cycle Assessment (LCA) can provide a quick tool to identify the life cycle hot-spots and focus research efforts to help minimize the burdens of a technology while maximizing its benefits. This talk examines the application of LCA to emerging nanotechnologies using case studies of nanosilver socks and a field emission display device (FED) containing carbon nanotubes (CNTs). The use of nanoscale silver in consumer products has exploded in popularity because of antimicrobial effects, making it the largest nanocomponent in use. The use of CNTs in nanoproducts is expected to offer numerous environmental benefits such as energy efficiency and the waste reduction. Although these products are considered beneficial to society, some attention must be given to the potential environmental impacts arising throughout the life cycle as production demands escalate.

See more of this Session: Life Cycle Studies Associated with Nanomaterials and Nano-Enhanced Products
See more of this Group/Topical: Topical C: Environmental Aspects, Applications, and Implications of Nanomaterials and Nanotechnology

(532a) Ultra-Rapid Elimination of Biofilms On Metal and Ceramic Surfaces Via the Combustion of a Nanoenergetic Coating

Shramik Sengupta[1], Rajagopalan Thiruvengadathan[2], Byung-Doo Lee[1], Brandon M. Smith[1], Sachidevi Puttaswamy[1], Keshab Gangopadhyay[3] and Shubhra Gangopadhyay[3], (1)Biological Engineering, University of Missouri, Columbia, MO, (2)University of Missouri Columbia, Columbia, MO, (3) Electrical and Computer Engineering, University of Missouri, Columbia, MO

In Heat Exchangers and Cooling-Water Systems, which are an integral part of a wide variety of industrial processes, the presence of biofilms can reduce the effective heat transfer coefficient by as much as 50%. Biofilms consist of microbials colonies embedded on the surface of the equipment, and protected by a layer of extra-cellular polymeric substances secreted by the same microbes.

Removal of the biofilms requires that the equipment be taken "offline" and subjected to chemical and physical treatments that not only take a long time, but also require the use of toxic chemicals. Reducing the down-time for the equipment and the development of processes that do not require the use and disposal of toxic chemicals, is thus highly desirable from both economic and environmental points of view.

We present here a novel process that uses nano-energetic materials to eliminate biofilms in just a few seconds. The process involves spray-coating of an approximately 100 μm thick layer of nano-energetic material on top of the biofilm, allowing it to dry, and igniting the dried coating to incinerate the biofilm. The nanoenergetic material used is a two component mixture consisting of aluminum nanoparticles (fuel) dispersed in a THV220A (fluoropolymer oxidizer) matrix. Upon ignition, the Al nanoparticles react with THV 220A exothermically, producing high temperatures (> 2500K) for an extremely brief period (~100mS) that destroys the biofilm underneath, along with the bacteria harbored within. However, due to the small mass of nanoparticles present (~1 mg/cm^2 of coating), the total *amount* of heat produced is low (~0.1 KJ/cm^2) and the underlying metal or ceramic surface does not suffer observable damages. Since the byproducts of the process consist only of carbonaceous residue (that can be easily washed away) and gaseous combustion products, the process is not extremely fast, but also environmentally friendly.

See more of this Session: Green and Renewable Energy Aspects of Nanotechnology
See more of this Group/Topical: Topical C: Environmental Aspects, Applications, and Implications of Nanomaterials and Nanotechnology

(532b) Life Cycle Energy Analysis and Midpoint Assessment of Multimegawatt Wind Turbines with Polymer Nanocomposite Blade Material

Laura A. Merugula[1], Bhavik R. Bakshi[1] and Vikas Khanna[2], (1)Lowrie Department of Chemical and Biomolecular Engineering, The Ohio State University, Columbus, OH, (2)Department of Civil and Environmental Engineering, University of Pittsburgh, Pittsburgh, PA

Life cycle analysis ofvapor-grown carbon nanofiber (VGCNF or CNF) manufacturing shows production ishighly energy-intensive with a cumulative energy demand that is influenced byfeedstock selection and dominated by processing energy needed to maintainreactor temperature for the catalytic pyrolysis of the hydrocarbon feed.However, at relatively low mass percentages, CNFs can reinforce polymercomposites yielding significant improvements in properties such as mode-1delamination resistance, vibrational damping, tensile strength, fatigueresistance and Young's modulus. A new method termed "pre-binding" has beendeveloped to reinforce polymer composites with CNFs at the interface of longfibers and resin. This method improves the manufacturability of large pieceswith CNF-reinforced polymer nanocomposites (PNCs) by overcoming processingissues inherent to previous methods of PNC fabrication. Materials scientistsare considering use of this emerging material for wind turbine blade productionto meet the demand for larger, more resilient utility-scale wind power plants.

Wind energy converters (WECs)on-line today are a relatively new technology with installations of multimegawattratings having been introduced to the modern market only in the last 15-20years. The largest 3-blade horizontal axis wind turbine (HAWT) rotor size inoperation is 126 meters in diameter with generator ratings from 5 to 7 MW whilethe growing market is dominated by new installations from 1 to 3 MW. Apparentlimits of fiberglass technology have been reached as rotor blades beyond about40 meters generally require some inclusion of carbon fiber to address stiffnessand edgewise fatigue, which become critical at this length. The less densecarbon fiber also leads to a positive deviation in trends of weight toincreasing size. Recent press releases communicate progress in R&D formeeting industrial-governmental collaborative goals for 20 MW plants possiblyby 2020 using available technologies with rotors expected to have diametersnear 200 meters.

In an even shorter timespan, numerous nanotechnology-enabled products have been introduced to theconsumer market with over 1,300 manufacturer-identifiedproducts populating The Project on Emerging Nanotechnologies consumer productinventory today. These products include automobile body parts with carbon nanotubes(CNTs) for enabling online electrostatic painting and a personal watercraftusing nanomaterials for reinforcement of the polymer composite for alightweight, stronger material.

While life cycle energyconsumption of CNF-reinforced PNCs for use in automobile body panels has beenevaluated, similar trends were not expected for life cycle energy in windturbines due to differences in operational phase energy consumption. Also,previous assessments of PNCs have not considered the method of pre-binding,which uses acetone as a vessel for dispersion and distribution of the CNFs ontoglass or carbon fiber matting.

Preliminary studies thatassumed replacement of polymer composite blade materials on a 2 MW and 5 MWplant with CNF pre-bound PNCs indicated a range of parameterization valuesunder which the energy return on investment (EROI) would not decrease. Thesestudies have been updated in a process-based life cycle assessment (LCA) toinclude more relevant process information and have been extended to a midpointanalysis using the CML and Traci methods to determine areas of focus foroptimization. An energy analysis was extended to a 5 MW WEC using cumulativeenergy demand and power generation figures from published LCA literature.

Impacts are reported perkilowatt-hour generation of electricity by the wind power plant and compared tothe base case, which is a 2 MW WEC in Middelgrund that is inventoried in the proprietaryEcoinvent database. In every feasible scenario, the EROI of wind power plants withthe PNC blade material remained competitive to thermoelectric power generation,and life cycle water consumption remained favorable. Where CNF production wasassumed to be based on benzene feedstock, energy analysis results were muchmore favorable than methane. Midpoint analysis, however, was assessed with methaneas the preferred hydrocarbon source for pyrolysis in nanofiber manufacture. Similartrends emerged under extension to the 5 MW system energy analysis.

Assuming a masscontribution of CNFs that ranges from 1 to 5 % by weight of each blade, theoverall mass of the nanomaterials is approximately 0.01-0.06% of the overallmass of the WEC system, yet there is direct correlation between the massfraction of the nanomaterials to nontrivial impacts per kWh compared to thebase case. Where studies have shown manufacturing of virgin nanomaterials tobe solvent-intensive, this study finds that the use phase of the nanomaterial,where CNFs are integrated into a product system, is highly solvent-intensive aswell. Current laboratory scale ratios of solvent spent to nanofibers dispersedare on the order of $5x10^2$ g solvent per g CNF. Maturation of thismaterial development must include a reduction of solvent use by optimization ofthe dispersion process and evaluation of solvent recovery systems for reusethat considers effects of nanomaterials in the operation. Photochemicaloxidation is a concern in terms of acetone emissions, the degree to which willbe dependent on recovery, reuse, and emissions controls. Also, the degree ofsolvent fresh feed required has an impact on the overall system's globalwarming potential and eutrophication potential due to the manufacturing phaseof the acetone.

Evaluating impacts of emergingtechnologies is complicated by the fast-pace of development and deployment,significant gaps in data and uncertainties about the future. Compilation oflife cycle data is unwieldy at best due to complex supply chains and regionalvariations in categories as energy supply and transportation. This is furthercomplicated by incomplete understanding of ecosystem service impacts, thoughwork is underway to quantify and compile ecosystem services for inclusion in LCA.System performance improvements will not occur with PNC integration inisolation of other technological changes within WEC design and operation, yet analysesas this are meant to give insight to areas of concern by identifying impactsand order of magnitude effects associated with this emerging material for theassumed use.

See more of this Session: Green and Renewable Energy Aspects of Nanotechnology
See more of this Group/Topical: Topical C: Environmental Aspects, Applications, and Implications of Nanomaterials and Nanotechnology

(532c) Advanced Oxidation Processes with Carbon Nanotubes: Surface-Promoted Formation of Hydroxyl Radical During Ozonation

Rebekah L. Oulton[1], Michael J. Nalbandian[1], Howard A. Fairbrother[2], Kevin Wepasnick[2] and David M. Cwiertny[1], (1)Chemical & Environmental Engineering, University of California, Riverside, Riverside, CA, (2)Department of Chemistry, Johns Hopkins University, Baltimore, MD

Advanced Oxidation Processes (AOPs), which utilize hydroxyl radical (•OH) to chemically degrade contaminants, are becoming increasingly utilized in water and wastewater treatment for disinfection and removal of organic micropollutants via chemical oxidation reactions. Catalytic ozonation represents an alternative AOP in which a solid substrate such as activated carbon or common metal oxides are used to promote •OH formation from ozone. This project explores the efficacy of using multi-walled carbon nanotubes (MWCNTs) as new materials for promoting •OH production during ozonation. Using para-chlorobenzoic acid (*p*-CBA) as a hydroxyl radical probe, aqueous suspensions of functionalized (i.e., surface oxidized) and nonfunctionalized MWCNTs increased •OH concentrations in bulk solution by an order of magnitude or more over levels observed with ozone alone and resulted in •OH levels comparable to more traditional ozone-based AOPs (e.g., hydrogen peroxide with ozone). Production of •OH was greatest on oxidized MWCNTs, and evidence suggests this enhanced •OH production cannot be explained solely by the increased dispersivity of oxidized MWCNTs in aqueous suspensions. Chemical routes for MWCNT functionalization were used to systematically vary the surface oxygen content of MWCNTs, and the ability of these functionalized materials to generate •OH was subsequently determined in batch ozonation experiments, again using *p*-CBA as a radical probe. Complementary surface analysis with X-ray photoelectron spectroscopy (XPS) combined with chemical derivatization was used to correlate •OH production not only to total MWCNT surface oxygen content, but also to the concentration of specific functional groups (hydroxyl, carbonyl and carboxylic acid groups) resulting from oxidation. Results to date show that •OH production increases with MWCNT surface oxidation, and that carboxyl surface groups appear largely responsible for this increase in •OH production.

See more of this Session: Green and Renewable Energy Aspects of Nanotechnology
See more of this Group/Topical: Topical C: Environmental Aspects, Applications, and Implications of Nanomaterials and Nanotechnology

(532d) Evaluating the Dynamics of An Integrated Electrokinetic and Zero-Valent Iron Nanoparticle System for Treatment of Hexavalent Chromium In Groundwater

Ryan Thacher and Massoud Pirbazari, Sonny Astani Department of Civil and Environmental Engineering, University of Southern California, Los Angeles, CA

Electrokinetic technology is a promising groundwater treatment strategy capable of transporting ionic contaminants through groundwater by application of an electric potential across a contaminated area. While this is effective in the transport and concentration of contaminants at an electrode, complimentary remedial action may still be required to enhance treatment of the contaminants. Zero-valent iron nanoparticles (nZVI) may potentially offer an ideal solution for this genre of *in-situ* environmental remediation. Their small size coupled with their high surface area to volume ratio as well as reductive capacity, make nZVI a seemingly ideal medium for subsurface remediation where other classical remediation strategies such as excavation are not possible. This may be in deep aquifers, urban areas, or under existing infrastructure.

A comprehensive understanding of the fate and transport of nZVI in the subsurface is just as critical to *in-situ* remediation as the high reactivity. Upon introduction to an aquifer environment, nZVI must be readily dispersible and remain stable so that they reach the targeted contamination zone. Several researchers have investigated transport of nZVI through porous media, however most of the published work is conducted in idealized batch studies. These investigations have indeed provided a good knowledge base, but limited information exists concerning the fate, transport, and toxicity of nZVI in the natural environment. The challenging issues at hand include: nanoparticle agglomeration, nZVI storage and age, components of soil matrices, aquifer hydraulic properties, presence of natural organic matter, pH, and redox potential, among others.

This study investigates the treatment of simulated groundwater containing hexavalent chromium (Cr^{VI}), a known carcinogen, in a dynamic system using combined electrokinetic and zero-valent iron nanoparticle technology. Continuous flow columns are charged with simulated natural soils, followed by saturation with solutions containing Cr^{VI}, variable humic acid concentration, and at multiple pH values. Upon saturation, electric potential is applied, and dilute solutions of nZVI are added at regular intervals for the entirety of the experiment.

This presentation will discuss the influence of natural aquifer characteristics such as pH, natural buffer, and natural organic matter content on the transport of nZVI through subsurface environments, and how this impacts treatment of contaminants. Furthermore, this research will evaluate the dynamics of coupling electrokinetic and nZVI technologies in decontamination of the subsurface environment.

See more of this Session: Green and Renewable Energy Aspects of Nanotechnology
See more of this Group/Topical: Topical C: Environmental Aspects, Applications, and Implications of Nanomaterials and Nanotechnology

(532e) Biodegradation of Carbazole by Microbial Cells Coated with Magnetite Nanoparticles

Jianmin Xing, National Key Lab of Biochemical Engineering, Institute of Process Engineering, Chinese Academy of Sciences, Beijing, China

Microbes can remove nitrogen from carbazole through a series of enzyme-catalyzed reactions. Microbial cells of *Klebsiella* sp. LSSE-H2 were coated with magnetic Fe_3O_4 nanoparticles, which were synthesized by a coprecipitation method followed by modification with ammonium oleate. The average size of the magnetic particles was found to be in the range from 10 to 15 nm. Transmission electron microscopy (TEM) cross section analysis of the cells showed further that the Fe_3O_4 nanoparticles were strongly absorbed by the surfaces of the cells and coated the cells. The coated cells had superparamagnetic properties. The coated cells not only had the same activity as free cells but could also be reused. Biodenitrogenation rate was increased with the addition of Tween 80. The activity decreased less than 23% after three recycles.

See more of this Session: Green and Renewable Energy Aspects of Nanotechnology
See more of this Group/Topical: Topical C: Environmental Aspects, Applications, and Implications of Nanomaterials and Nanotechnology

(586a) Comparison of Photoreactor Designs for Oxidation of Dilute Aqueous Waste Contaminants

Amanda M. Grannas[1], **Dorothy W. Skaf**[2], Kevin C. Brodwater[2] and Montanna Herdemian[2], (1) Chemistry, Villanova University, Villanova, PA, (2)Chemical Engineering, Villanova University, Villanova, PA

The viability of photocatalytic reactions in advanced oxidation processes for removal of dilute aqueous contaminants was evaluated using a variety of reaction systems. Using acetaminophen as analyte, the oxidation reaction was studied using various operating conditions and reactor designs to better understand how the reaction conditions impact conversion and potentially, product selectivity. Catalyst concentration, reagent concentration, metal loading on the catalyst, oxygen purging, solution pH, solution temperature, addition of oxidant, batch versus continuous operation, light intensity and semiconductor stabilization were investigated. Literature results for reaction intermediates and reaction kinetics were used to interpret experimental results. A main focus was optimization of reaction conditions that can make photocatalytic oxidation of dilute aqueous contaminants more sustainable and economical. Catalyst stability, consumption of additional reagents, value and/or safety of the degradation products and energy consumption were considered for the various reaction options.

See more of this Session: Environmental Applications of Nanotechnology and Nanomaterials
See more of this Group/Topical: Topical C: Environmental Aspects, Applications, and Implications of Nanomaterials and Nanotechnology

Study of the Photodeposition of Noble Metals on BiOCl for the Photocatalytic Decomposition of Rhodamine B

Liang Kong, Zheng Jiang, Henry H.-C. Lai, Rebecca Nicholls, Tiancun Xiao, Peter P. Edwards**

Department of Chemistry, Inorganic Chemistry Laboratory, University of Oxford, U.K.

Introduction

Extensive efforts have been made to discover robust semiconductor photocatalysts for efficient utilization of the most sustainable solar energy in photocatalytic water splitting, CO_2 reduction and environment cleanup[1-4]. The p-type BiOCl has been recently emerged as an active photocatalysts only under the illustration of UV light which accounts for approximately 5% energy of sunlight spectrum[5]. This photocatalyst is still facing the similar major challenges encountered by most photocatalysts: catalyst stability and energy loss associated with the recombination of photogenerated charge carriers (electrons and holes). A variety of strategies, such as the formation of heterojunctions[6], impurity doping and surface metallisation, have been attempted to extend the harvesting of sunlight into visible light region[7]. Surface metallisation using noble metals is one of the most effective methodologies for trapping photogenerated electrons due to the high conductivity, excellent electron-trap capability and additional plasmonic effect of the nano-sized noble metals. To date, the semiconductor surface metallisation mainly concentrated on the n-type semiconductors[8] rather than p-type semiconductor photocatalysts. Most recently, Ag/BiOBr and Pt/BiOI have demonstrated enhanced photocatalytic performance[9,10], whereas, the effects of noble metals deposition on the photocatalytic performance BiOCl has not been explored.

In this work, we deposited a small amount of various noble metals on BiOCl *via* a simple photo-deposition method and found that the noble metals have different influences on the structure, properties and photocatalytic performance of BiOCl. The interaction between noble metal and BiOCl has been discussed to interpret the effects of surface-deposited noble metal.

Results

Crystallinity and morphology of the metal/BiOCl (here after NM/BOC)

The XRD patterns (Fig. 1A) of the NM/BOC samples are well consistent with the tetragonal BiOCl phase (JCPDS no. 01-082-0485[11]) without obvious change of the characteristic diffraction peaks, suggesting the surface deposition of noble metals does not affect the crystal structure of BiOCl. There is no diffraction peaks of noble metals observed, which can be attributed to the low loading (0.5 wt%) and the high dispersion of noble metals on the surface of BiOCl. Although the lattice parameters of the BiOCl are not affected significantly, as shown in Table 1, the deposited noble metal species drastically decrease the crystallite size in following order of BiOCl > Rh/BOC > Pt/BOC > Pd/BOC > Ag/BOC.

Table 1. Crystal parameters, band-gap and photodegradation rate constant of NM/BOC

Sample ID	Crystallite size (Å)	Lattice Parameters		k_{RhB} (min^{-1})
		a (Å)	*c* (Å)	
BiOCl	451	3.892	7.376	0.247
0.5 wt% Pt-BiOCl	355	3.889	7.361	0.169
0.5 wt% Pd-BiOCl	325	3.887	7.365	0.194
0.5 wt% Rh-BiOCl	384	3.893	7.383	0.140
0.5 wt% Ag-BiOCl	292	3.888	7.387	0.253

As shown in Fig.1(B), Ag/BiOCl sample remains the flake-like morphology as same as the pure BiOCl (SEM not shown here). The average diameter of the flakes is approximately 200 nm. The TEM image (Fig. 1C)of the Ag/BiOCl sample further evidences its flake-like morphology and the nano-sized metal Ag particles in diameter approximately 5nm were highly dispersed on the surface of the BiOCl flakes in agreement with the XRD results. Other NM/BiOCl samples have the similar flake morphologies.

Fig. 2 (A)XRD patterns of as-syntheisised undoped BiOCl and samples with 0.5 wt% photodeposited Pt, Pd, Ag and Rh, (B)SEM and (C)TEM image of 0.5 wt% Ag/BiOCl sample

XPS analysis

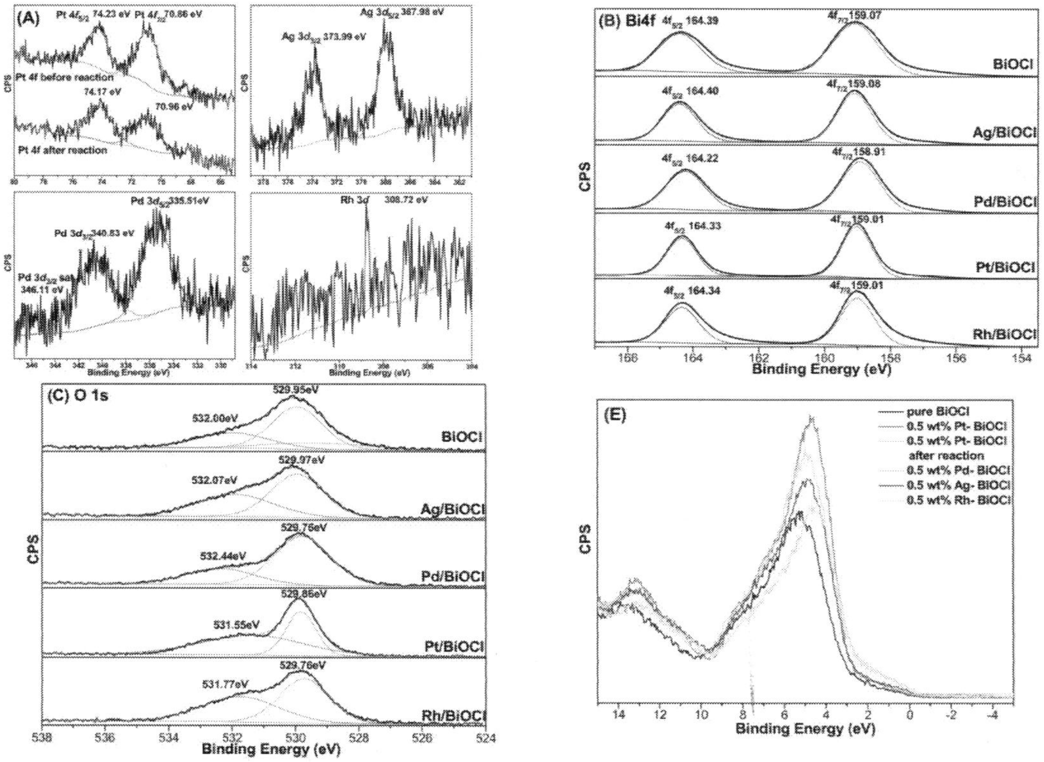

Fig.2 XPS spectra of (A) Bi 4*f*, (B) O 1*s*, (C) Pt 4*f* (before and after photocatalytic reaction), Ag 3*d*, Pd 3*d* and Rh 3*d*, and (E) VBM of all the NM/BiOCl samples

The characteristic binding energies (B.E.) of Bi, Cl, O and noble metals can be well identified from the XPS results of the corresponding samples. The XPS-derived surface atomic ratios of noble metal, Cl, O and Bi elements the samples are listed in Table 2. These results confirm the desired compositions of the BiOCl and NM/BOC samples. The noble metals in the study are all in metallic states as shown in Fig 2(A) without any signals associated with noble metal cations observed, suggesting the photoreduction is a successful strategy in deposition of noble metals on the BiOCl photocatalyst. Fig 2(B) shows the Bi $4f_{5/2}$ and Bi $4f_{7/2}$ XPS core lines with B.E located around 164.3 eV and 159.0 eV, revealing the Bi in BiOCl is Bi^{3+} [12]. The Bi 4*f* core lines shifted towards lower energies for the NM/BOC samples except for Ag/BOC whose Bi 4*f* core lines increased about 0.2 eV. The most significant red-shifts of Bi 4*f* B.E. are observed on the Pd/BOC sample. The detailed B.E. changes may be found in Table 3. There is no obvious changes can be observed from Cl $2p_{1/2}$ (199.4 eV) and Cl $2p_{3/2}$ (197.8 eV) peaks of Cl^- in the BiOCl and NM/BOC, suggesting the supported noble metal can not affect the inter-layered Cl^- [13,14]. There are two different oxygen species presented on the surface of the series samples (Fig. 2C). The O 1*s* B.E. around 532.0 eV is assigned to the Bi-O bond in BiOCl, while the O 1*s* peak at 529.9 eV due to the surface hydroxyl groups of the sample[15,16]. Moreover, the O 1s of Bi-O shift towards higher energy against the trend of Bi 4f on the NM/BOC samples except for that of Ag/BOC, which further confirms that surface metallisation affect the chemical environment of the composite. The valence band XPS of the samples are compared in Fig.2D, in which the shifts of VB edges (VBMs) towards lower energy on the NM/BOC than VB of BOC are clearly observed and the shifts significantly depend on the noble metal specials.

Table 2. Surface elemental composition (%) of the NM-BiOCl samples

Atomic Ratio	Bi	O	Cl	NM
BiOCl	39.2	38.2	22.6	-
Rh-BiOCl	36.3	52.4	10.9	0.49
Pd-BiOCl	37.4	36.1	25.9	0.56
Pt-BiOCl	38.3	34.2	27.3	0.18
Pt-BiOCl after reaction	33.4	43.8	22.7	0.19
Ag-BiOCl	36.8	41.9	20.9	0.46

Table 3. XPS characteristic peak positions of the NM-BiOCl samples

Sample ID	Peak position (eV)				
	Bi $4f_{5/2}$	Bi $4f_{7/2}$	Cl $2p_{1/2}$	Cl $2p_{3/2}$	O 1*s*
BiOCl	164.4	159	199.5	197.8	530.1
Pt-BiOCl	164.2	158.9	199.3	197.7	529.8
Pt-BiOCl after reaction	164.3	159.0	199.4	197.8	529.9
Pd-BiOCl	164.2	158.9	199.3	197.7	529.7
Rh-BiOCl	164.3	159.0	199.3	197.7	529.8
Ag-BiOCl	164.4	159.1	199.6	197.9	530.1

Uv-Vis Spectra

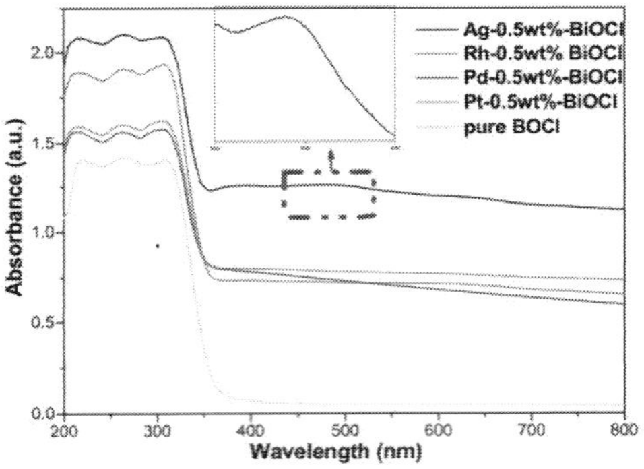

Fig. 3 UV-vis diffuse reflectance spectra of NM- BOC photocatalysts

Fig. 3 shows the UV/vis diffuse reflectance spectra of BiOCl and NM/BOC samples. T BiOCl can only absorb UV light with wavelength below 360 nm, while all the NM/BOC samples can harvest light through the whole UV-Vis region, suggesting even small amount of noble metals can greatly improve the light absorption in comparison to pure BiOCl. The enhanced absorption of NM/BiOCl samples both in UV and Visible regions may be attributed to the close field surface plasmonic effect (or the scattering effect)[9]. The Ag/BOC sample exhibits the strongest plasmonic effect among the NM/BOC samples. The indirect and direct band-gap energy of NM/BOC can thus be estimated from the data plot of $(\alpha \cdot hv)^{1/2}$ and $(\alpha \cdot hv)^2$ *versus* photo energy (hv) [5] and listed in Table 2, from which the introducing of noble metals decrease both the indirect band-gap and direct band-gap.

Photocatalytic activity

Photocatalytic decomposition of RhB, a chemically stable organic dye with a characteristic absorption peak at 554 nm, was chosen as a probe reaction to clarify the effects of deposited noble metals on the BiOCl. As shown in Fig. 4(A) and (B), all the samples can completely decompose Rh.B within 30 minutes UV light irradiation under the same reaction conditions. The photocatalytic activity highly dependes on the type of noble metal, with the activity changes in the order of Ag/BOC > pure BiOCl > Pd/BOC > Pt/BOC > Rh/BOC. The results suggest that the modification of noble metals cannot enhance the photocatalytic activity of the BiOCl samples even though they have enhanced absorption of irradiation.

Although most of the NM/BOC are less active than BOC except for Ag/BOC, the NM/BiOCl samples showed higher stability after UV-light irradiation. For example, the fresh and used 0.5 wt% Pt/BiOCl samples shows the very similar XRD patterns (Fig.4C) except for the stronger Bragg diffraction intentisty for the sample post photocatalysis. The enhanced intensity may be attributed to the growth of the crystallite during the photocatalytic reaction. The XPS spectra of fresh and used Pt/BiOCl samples were also compared, as shown in Table 2, indicating there is no obvious difference between the samples before and after the photocatalytic reaction. Both the XRD pattern and XPS spectra of the samples after reaction confirmed that the metal-modified samples are considerably stable under UV light irradiation.

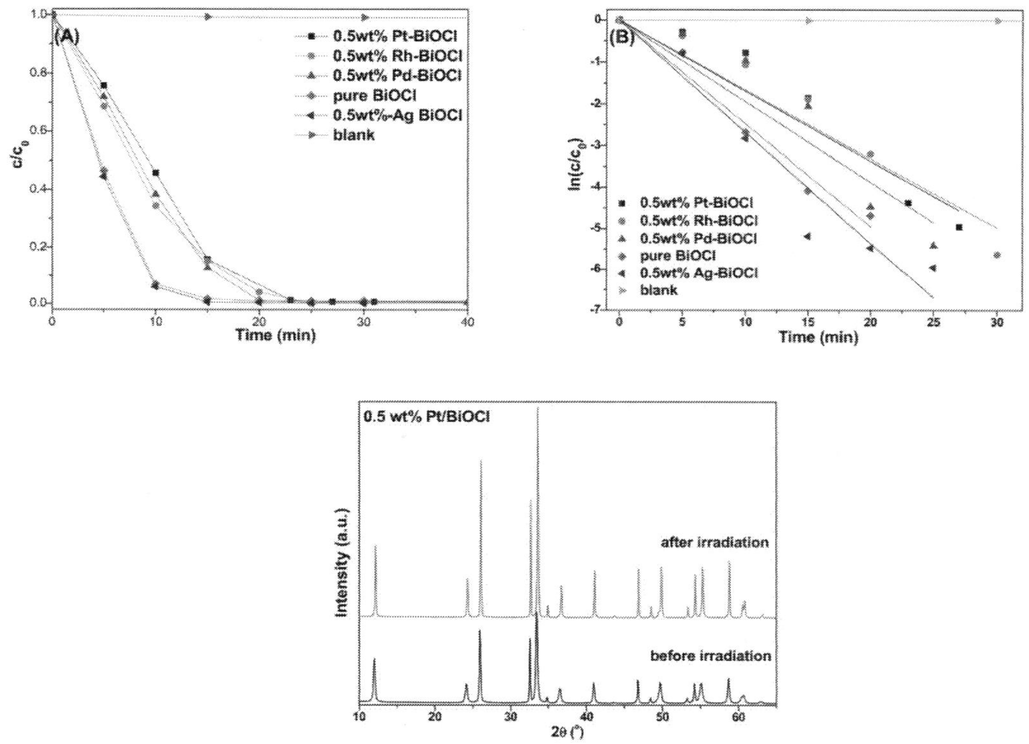

Fig. 4 (A) Temporal course of the degradation of RhB in the presence of different noble metal- BiOCl; (B) Kinetic linear curves of RhB phpotocatalytic degrdation over noble metal- BiOCl and (C) XRD patterns of 0.5 wt% Pt- BiOCl sample before and after UV-light irradiation

Working mechanism

Understanding the reason of the superior photocatalytic activity and the working mechanism of the Nm/BiOCl composite catalysts is essential for fundamental investigation of semiconductor photocatalysts. The formation of Schottky barrier between metals and semiconductor could enhance the photocatalytic activity due to their different Fermi levels, characterised by the work function of the metals and the band structure of the semiconductors[8]. Photoexcited electrons in the conduction band of the semiconductor will raise its Fermi level to more negative values[17], usually leading to enhanced photo-activity of most photocatlysts, such as TiO_2, while our experimental results differ from previous reports that the modification of noble metals decrease the activity of photocatalysts. According to the different work functions of the noble metals, we obtain the height of the noble metals in the reducing order of Pd > Pt > Ag > Rh. This trend is in accordance with the photocatalytic activity consequence of NM/BiOCl samples, i.e., the noble metal with higher Fermi level shows better photocatlaytic activity. However, the work functions of the NM are more positive than the CB of BiOCl, which leads to photogenerated electrons from the VB easily flow into noble metal rather than the CB of BOC and decreased photocatalytic activity of the NM/BOC due to the less opportunity to generate reactive $O_2^{-}\bullet$. At the same time, the chemical potential of the photogenerated holes are less oxidative due to the shift of VB towards to more negative positions, which means the decreased oxidative capability of VB holes to oxidize the dye molecules. The overall results are decreased activity of the NM/BOC in comparison with pure BiOCl.

Conclusion

Noble metal-modified BiOCl photocatalysts have been synthesised *via* a two-step precipitation-photodeposition method. The metallic state modification on the surface of BiOCl significantly enhances the absorbance in the visible-light region, *i.e.* decreases the band-gap to about 2.7 eV, however, the modification cannot effectively enhance their photocatalytic activity with respect to the pure BiOCl in photodecomposion of RhB. The decreased photocatalytic activity of the NM/BOC can be well interpret by the work function difference between the NMs and BiOCl.

Acknowledgement

This work was financially supported by the principal's major fund at Jesus College, University of Oxford, Oxford Challenge Seed Fund, UK Photocatalysis Network, UKSHEC (EPSRC). Dr Jiang appreciates John Houghton Fellowship at Jesus College (Oxford) and the grants of Royal Society for international conference (TG092414) and international collaboration (TG101750).

Reference

(1)Ohtani, B. *Journal of Photochemistry and Photobiology C: Photochemistry Reviews* **2010**, *11*, 157-178.

(2)Jiang, Z.; Yang, F.; Yang, G.; Kong, L.; Jones, M. O.; Xiao, T.; Edwards, P. P. *Journal of Photochemistry and Photobiology A: Chemistry* **2010**, *212*, 8-13.

(3)Jiang, Z.; Al-Shahrani, F.; Lin, T.-W.; Cui, Y.; Xiao, T. In *Studies in Surface Science and Catalysis*; Dongyuan Zhao, S. Q. Y. T., Chengzhong, Y., Eds.; Elsevier, 2007; Vol. Volume 165; pp 355-359.

(4)Jiang, Z.; Xiao, T.; Kuznetsov, V. L.; Edwards, P. P. *Philosophical Transactions of the Royal Society A: Mathematical, Physical and Engineering Sciences* **2010**, *368*, 3343-3364.

(5)Zhang, K.-L.; Liu, C.-M.; Huang, F.-Q.; Zheng, C.; Wang, W.-D. *Applied Catalysis B: Environmental* **2006**, *68*, 125-129.

(6)Shamaila, S.; Sajjad, A. K. L.; Chen, F.; Zhang, J. *Journal of Colloid and Interface Science* **2011**, *356*, 465-472.

(7)Kong, L.; Jiang, Z.; Xiao, T.; Lu, L.; Jones, M. O.; Edwards, P. P. *Chemical Communications* **2011**, *47*, 5512-5514.

(8)Ohtani, B.; Kakimoto, M.; Nishimoto, S.; Kagiya, T. *Journal of Photochemistry and Photobiology A: Chemistry* **1993**, *70*, 265-272.

(9)Yu, C.; Fan, C.; Meng, X.; Yang, K.; Cao, F.; Li, X. *Reaction Kinetics, Mechanisms and Catalysis* **2011**, *103*, 141-151.

(10)Yu, C.; Yu, J. C.; Fan, C.; Wen, H.; Hu, S. *Materials Science and Engineering: B* **2010**, *166*, 213-219.

(11)Keramidas, K. G.; Voutsas, G. P.; Rentzeperis, P. I. *Zeitschrift fur Kristallographie* **1993**, *205*, 35-40.

(12)Xiao, Q.; Zhang, J.; Xiao, C.; Tan, X. *Catalysis Communications* **2008**, *9*, 1247-1253.

(13)Ge, L. *Journal of Molecular Catalysis A: Chemical* **2008**, *282*, 62-66.

(14)Drelinkiewicz, A.; Sobczak, J. W.; Sobczak, E.; Krawczyk, M.; Zieba, A.; Waksmundzka-Góra, A. *Materials Chemistry and Physics* **2009**, *114*, 763-773.

(15)V.I., N.; D., G.; Dzhurinskii, B. F.; N.P., S.; Y.V., S. *Zh. Neorg. Khimii* **1975**, *20*, 2307.

(16)Martensson, N.; Malmquist, P. A.; Svensson, S.; Basilier, E.; Pireaux, J. J.; Gelius, U.; Siegbahn, K. *Nouveau, J. Chim.* **1977**, *1*, 191

(17)Subramanian, V.; Wolf, E. E.; Kamat, P. V. *Journal of the American Chemical Society* **2004**, *126*, 4943-4950.

(586c) Synthesis and Characterization of Magnetic Nanoparticles for Enhanced Gas-Liquid Mass Transfer

Alexander P. Mathews, Department of Civil Engineering, Kansas State University, Manhattan, KS, Dambar B. Hamal, Department of Chemistry, Kansas State University, Manhattan, KS, Paul Owings, Civil Engineering, Kansas State University, Manhattan, KS and Ken J. Klabunde, Chemistry, Kansas State University, Manhattan, KY

Abstract

Magnetite particles weresynthesized by the coprecipitation of iron (II) and iron (III) chlorides in thepresence of a surfactant and ammine base. These particles had relatively lowinternal surface areas, but high external surface due to the small particlediameter. Silica coated magnetite particles were synthesized by the hydrolysisof tetraorthosilicate (TEOS) and the coprecipitation of iron chlorides in thepresence of different surfactants and an ammine base. Commercial magnetitenanoparticles were also coated with silica using TEOS and a base in thepresence or absence of a dispersant. All samples were characterized using XRD,SEM, TEM, and nitrogen adsorption to obtain diffraction patterns, particlemorphology data, and BET surface areas. Adsorption isotherm studies wereconducted for the adsorption of trichloroethylene from aqueous samples on tolow surface area magnetite nanoparticles and high surface area silica coatedmagnetite nanoparticles. Studies were also conducted on the effect of nanoparticleson the effect of high and low surface area nanoparticles on gas-liquid masstransfer in aqueous systems. This paper will present data on the synthesis andcharacterization of these nanoparticles and their potential to enhance masstransfer in gas-liquid systems.

See more of this Session: Environmental Applications of Nanotechnology and Nanomaterials
See more of this Group/Topical: Topical C: Environmental Aspects, Applications, and Implications of Nanomaterials and Nanotechnology

(586d) Visible-Light-Driven Photodegradation of Contaminants In Water Over Surface-Engineered BiOBr Semiconductor Micro/Nano-Structures

Zheng Jiang[1], Liang Kong[2], Tiancun Xiao[2] and Peter P. Edwards[2], (1)Jesus College, University of Oxford, Oxford, United Kingdom, (2)Department of Chemistry, Inorganic Chemistry Laboratory, University of Oxford, Oxford, United Kingdom

Visible-Light-Driven Photodegradation of Contaminants in Water over Surface-Engineered BiOBr Semiconductor Micro/Nano-Structures

Zheng Jiang[1,2], Liang Kong[1], Tiancun Xiao[1], Peter P. Edwards[1]

(1)Jesus College, University of Oxford, Oxford, OX1 3DW, U.K.; (2)Department of Chemistry, Inorganic Chemistry Laboratory, University of Oxford, Oxford, OX1 3QR, U.K.

Visible-Light-Driven Photocatalytic degradation of contaminants (PDC) over semiconductor photocatalysts represents an advanced, cost-efficient and promising water purification technology because it can utilize the large portion of energy in the free sun light[1,2]. However, the implementation of PDC technique is is still facing a great challenge over the effective and stable visible-light-responsive photocatalysts with desired structure and surface functionality. Extensive efforts have been made to enhance the adsorption of sunlight and depress the recombination of the photogenerated charges for maximum photocatalytic efficiency. Semiconductor BiOBr is an attractive photocatalyst with excellent Uv-Vis light illumination despite its adsorption of sunlight is limited by its relatively large bandgap[3,4]. This talk will focus on the strategies we developed to resolve the crucial issues over the BiOBr semiconductor by semiconductor surface-engineering. An efficient and controllable synthesis methodology has been developed to prepare BiOBr mico-/nano-structures with desired visible-light adsorption.

First, a rapid synthesis route was developed to prepare BiOBr nanoflowers with exposed highly reactive surfaces and it was further adopted to prepare heterojucntions composed of BiOBr and second semiconductors. The results reveal the surface-oriented BiOBr and its heterojunctions possess hierarchical architectures and ultrahigh activity under visible-light irradiation in photodegradation of several typical refractory model organic pollutants. The synergism effects of the enhanced surface absorption, enhanced light adsorption, mitigated light-scattering, and special hierarchical structures were unearthed through key characterizations. Second, visible-light-induced photodeposition was adopted to prepare the surface supported noble-metal/BiOBr photocatalysts and the effects of the NM on the BiOBr surface were explored. It was found that the surface NMs deteriorated the photocatalytic performance of the BiOBr photocatalysts, although the surface NMs significantly improve the stability of the BiOBr under long-term light exposure. The mechanism was proposed on basis of the surface engineering, and band structure of the multi-structures.

The correlation of the photocatalytic activity to its structure and surface will be discussed on basis of the detailed physicochemical characterization and theoretical analyses.

References:

1. Z Jiang, T Xiao,V Kuznetzov, P.P. Edwards, ***Phil. Trans. R. Soc. A***; 2010, 368; 3343-3364;

2. Z. Jiang, F. Yangb, N. Luo, B. T. T. Chu, D. Sun, H. Shia, T. Xiao, P. P. Edwards, ***Chem. Commun.***, 2008, 6372-6374

3. L. Kong, Z. Jiang, T. Xiao, L. Lu, M. O. Jones and P. P. Edwards, ***Chem. Commun.***, 2011, 47, 5512-5514

4. Z. Jiang, F. Yang, G. Yang, L. Kong, M. O. Jones, T. Xiao, P. P. Edwards, ***J PHOTOCHEM PHOTOBIOL A-CHEM*** , 2010,212(1): 8-13.

See more of this Session: Environmental Applications of Nanotechnology and Nanomaterials
See more of this Group/Topical: Topical C: Environmental Aspects, Applications, and Implications of Nanomaterials and Nanotechnology

(586e) Magnetic Catch & Release: Reversible Organic Contaminant Adsorption and Enrichment From Water

Roland Fuhrer, Inge K. Herrmann, Evagelos K. Athanassiou, Robert N. Grass and Wendelin J. Stark, Institute for Chemical- and Bioengineering, ETH Zurich, 8093 Zurich, Switzerland

There is an increasingneed for fast and efficient extraction methods to remove low concentrated (ppblevels) organic compounds from water with high adsorbent recyclability. Surface-modified magnetic nanoparticles can be used in extraction processesas they readily disperse in common solvents and combine high saturationmagnetization with excellent accessibility. Reversible and recyclableadsorption and desorption through solvent changes and magnetic separationprovide technically attractive alternatives to classical solvent extraction. Carbon-coatedcobalt nanoparticles with a thin polymer layer were tagged withβ-cyclodextrin. The resulting material reversibly adsorbed organiccontaminants in water within minutes. Isolation of the immobilized inclusioncomplex was easily carried out within seconds by magnetic separation due to thestrong magnetization of the nanomagnets (metal core). The trapped moleculeswere fully and rapidly recovered by filling the cyclodextrin cavity with amicrobiologically well accepted substitute, e.g. benzyl alcohol.Phenolphthalein was used as a model compound for organic contaminants such aspolychlorinated dibenzodioxins (PCDDs) or bisphenol A (BPA). The nanomagnetscan be regenerated under mild conditions and reused in the next cycle at fullefficiency (16 recycles tested). Tagging cyclodextrins with magnetic, stable nanoparticles(carbon shell, polymer layer) makes them magnetoresponsive and may lead to anew generation of adsorbents in separation, contaminant enrichment or drugdelivery. Experiments at ultralow concentrations (160 ppb) underline the highpotential of cyclodextrin modified nanomagnets as a fast, recyclable extractionmethod for organic contaminants in large water streams or as an enrichment toolfor analytics.

Reference:

1) R. Fuhrer, I.K. Herrmann, E.K.Athanassiou, R.N. Grass, W.J. Stark, **Langmuir**,2011, 27(5), 1924-9

Figure:

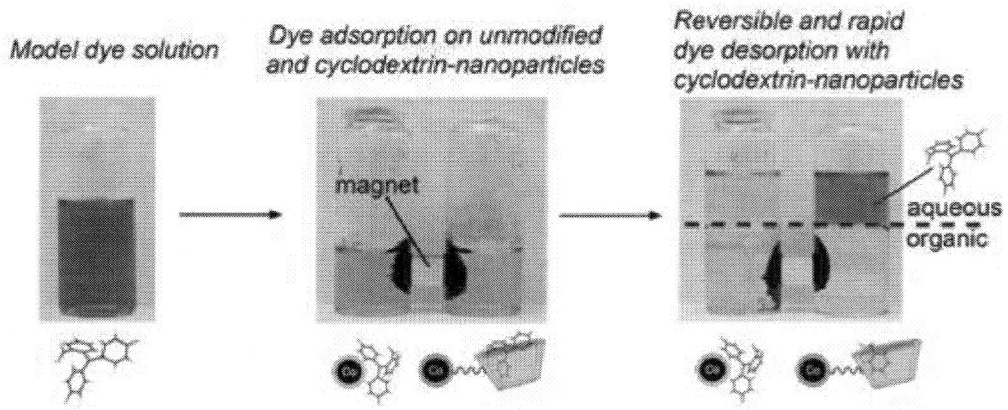

See more of this Session: Environmental Applications of Nanotechnology and Nanomaterials
See more of this Group/Topical: Topical C: Environmental Aspects, Applications, and Implications of Nanomaterials and Nanotechnology

(586f) Improved Photocatalytic Activity Under Visible Light Irradiation of Nanosized-TiO2 Co-Doped with Vanadium and Nitrogen

Renuka Jaiswal[1], Rupali Dholam[1], Nainesh Patel[2], Antonio Miotello[2] and **Dushyant C. Kothari**[1], (1) Centre for Nanosciences & Nanotechnology, University of Mumbai,, Mumabi, India, (2)Physics, Universita degli Studi di Trento, Povo (Trento), Italy

Heterogeneous nanocrystalline titania (TiO_2) has been extensively investigated due to its notable photocatalytic capability for the industrial waste water treatment. However, an obvious problem hindering the application of TiO_2 is its poor response to visible light due to its wide bandgap. Herein, TiO_2 photo-catalyst is synthesized by co-doping vanadium and nitrogen by sol-gel method to sensitize TiO_2 in visible light. The concentrations of V and N were varied and optimized to enhance photo-catalytic activity. In order to increase the active surface area, the crystalline powder was ball milled to nanosize. The photo-catalysts were characterized by XRD, FT-IR, SEM, EDS, BET surface area, XPS, and UV–vis diffuse reflectance spectra. The co-doped TiO_2 nanocrystal has a narrower band gap than singly V-doped TiO_2 and singly N-doped TiO_2. The XPS detected vanadium (V^{4+} and V^{5+}) and nitrogen ions on the surface of photo-catalysts, but no diffractive peak of vanadium oxide or titanium nitride was observed in XRD spectra. High energy ball-milling was able to increase the surface area due to reduction in particle size below 50 nm as confirmed by SEM. The co-doped TiO_2 photo-catalyst showed significantly better photocatalytic activity for the degradation of organic dyes such as Rhodamine B and methylene blue under visible light irradiation ($\lambda > 420$ nm) as compared with V-doped TiO_2 and N-doped TiO_2. The incorporation of N and V in TiO_2 lattice induces the appearance of passivated shallow acceptor and donor levels near the valence and conduction band, respectively, causing narrowing of the band gap. In addition, these levels can also act as traps for photoexcited holes or electrons to reduce the recombination between photo-generated charges thus contributing to enhance the photocatalytic activity.

See more of this Session: Environmental Applications of Nanotechnology and Nanomaterials
See more of this Group/Topical: Topical C: Environmental Aspects, Applications, and Implications of Nanomaterials and Nanotechnology

AUTHOR INDEX

AUTHOR INDEX